BIOLOGY BASIC FACTS

T A McCahill BSc DipEd MIBiol

HarperCollins*Publishers*

HarperCollins Publishers
PO Box, Glasgow G4 0NB, Scotland

First Published 1982
Fourth edition 1996

Reprint 10 9 8 7 6 5 4 3 2 1

ISBN 0 00 470994 2

Printed and bound in Great Britain by
Caledonian International Book Manufacturing Ltd,
Glasgow, G64

Introduction

Collins Gem *Basic Facts* is a series of illustrated GEM dictionaries in important school subjects. This new edition has been revised and updated to widen the coverage of the subject and to reflect recent changes in the way it is taught in the classroom and in the content of exam syllabuses.

Bold words in an entry identify key terms which are explained in greater detail in entries of their own; important terms which do not have separate entries are shown in *italic* and are explained in the entry in which they occur.

Other titles in the series include:

Gem *Business Studies Basic Facts*
Gem *Chemistry Basic Facts*
Gem *Computers Basic Facts*
Gem *Geography Basic Facts*
Gem *History Basic Facts*
Gem *Mathematics Basic Facts*
Gem *Physics Basic Facts*
Gem *Science Basic Facts*
Gem *Technology Basic Facts*

abdomen **1.** (in mammals) The part of the body separated from the **thorax** by the diaphragm, containing **stomach, liver, intestines,** etc.
2. (in insects) The posterior third region of the body.

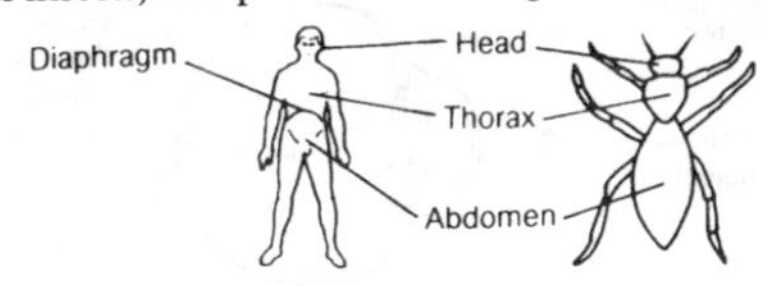

abdomen The position of the abdomen in a human being and an insect.

abiotic factor See environment.

abscission The shedding of **leaves, fruit,** and unfertilized **flowers** from plants by the formation of a layer of cork **cells** which seal the plant surface and eventually cut off food and water from the part to be shed.

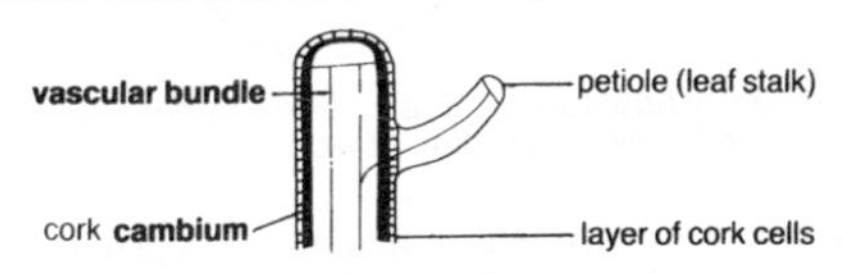

abcission Cork cells seal off a leaf.

absorption (of food) The process by which digested food particles pass from the **gut** into the bloodstream. In mammals absorption occurs in the **ileum.**

accommodation The way in which the **eye** of mammals can change its sharp focus from near to distant objects, and vice versa, by means of contraction or relaxation of the **ciliary muscles,** so altering the shape and hence the focusing properties of the **lens.**

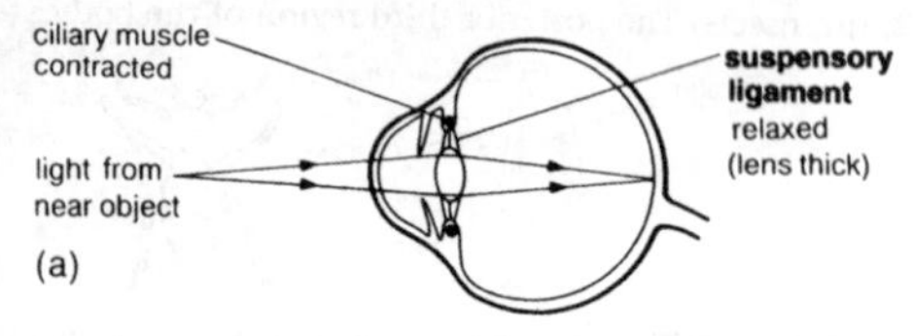

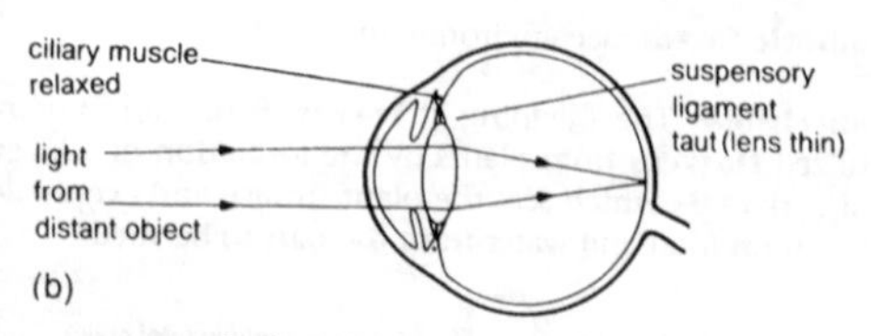

accommodation (a) Eye focused on near object. (b) Eye focused on distant object.

acid A substance which releases hydrogen **ions** in **water** and has a **pH** less than 7. An example is hydrochloric acid in the human **stomach.**

acid rain Rain which contains dissolved **acid** and is therefore a source of **pollution.** The acidity is caused mainly by the release of **sulphur dioxide** when **fossil fuels** are burned. The sulphur dioxide reacts with water in the atmosphere to form **sulphuric acid** which corrodes buildings, lowers the **pH** of soil and lakes, and affects plant and animal life.

active transport The movement of materials against a *concentration gradient* (see **diffusion**) using **metabolic energy.**
Examples are (a) the uptake of **mineral salts** from **soil** by plant **root hairs** and (b) the reabsorption of certain substances by the mammalian **kidney.**

adaptation The development of structures within organisms so that they are more efficiently adapted to their **environment,** e.g. plant **leaves** are constructed in such a way as to contribute to the efficiency of **photosynthesis.**
Leaves are
(a) thin, allowing rapid **diffusion** of air, thus facilitating **gas exchange;**
(b) flat and present a large surface area to the light.
The **cells** immediately beneath the upper **epidermis** are called **palisade mesophyll** cells. They are closely packed and contain many **chloroplasts,** most of which are situated at the upper part of the cell, thus obtaining maximum light. Between the palisade cells and the lower epidermis are the **spongy mesophyll** cells, which contain fewer chloroplasts. These cells are loosely packed, being separated by air spaces, which allow free circulation of air between the leaves and the atmosphere via the **stomata.**
The positioning of most of the stomata on the lower epidermis means that the important palisade layer is not interrupted by too many air spaces.
Water for photosynthesis is transported in **xylem** ves-

sels in the mid-rib and veins, while the synthesized **carbohydrate** is transported throughout the plant via the **phloem** sieve tubes.

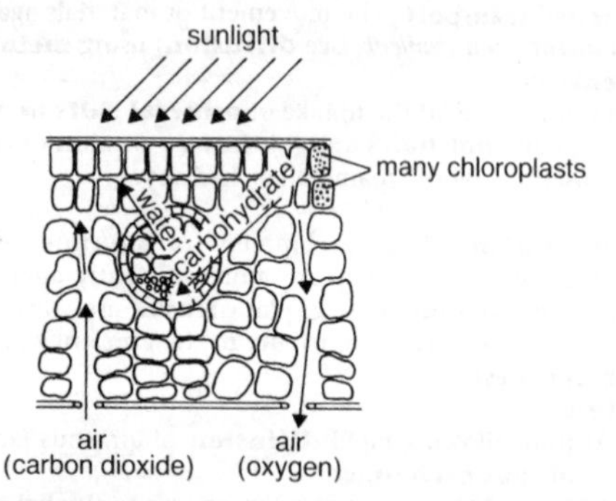

adaptation The structure of a leaf is adapted for efficient photosynthesis.

ADH See **antidiuretic hormone.**

adipose tissue Mammalian tissue consisting of **fat** storage **cells,** located under the **skin,** around the **kidneys,** etc.

adolescence The period in the human **life cycle** between **puberty** and maturity.

ADP See **ATP.**

adrenal glands A pair of **endocrine glands** situated

anterior to the mammalian **kidneys** and secreting the **hormone** *adrenalin* which causes increased heart-beat, breathing, etc., in response to conditions of stress.

aerobe An organism which requires oxygen in order to survive. See **anaerobe, respiration**.

aerobic bacteria See **sewage disposal**.

agar A jelly obtained from seaweeds which is used as a medium for culturing **bacteria**. A food source, e.g. **glucose**, is added and the liquid agar is poured into a Petri dish where it solidifies. Bacterial sources are added to this *nutrient agar plate* and after two days at a suitable temperature (usually 37 °C), the bacterial colonies become visible as a result of rapid cell division. See diagram.

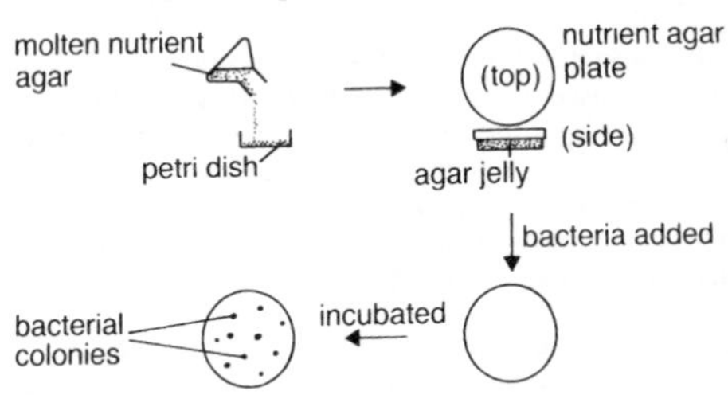

agar Bacteria are cultivated on a nutrient agar plate.

agglutination The process by which **red blood cells** clump together when the *antigens* on their surfaces react with complementary **antibodies**. See **blood groups, blood transfusion**.

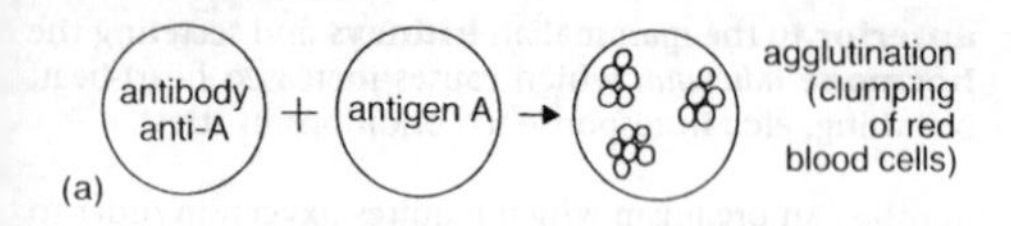

agglutination

agriculture The practice of farming, involving the cultivation of **soil**, crop production and raising livestock.

AIDS (aquired immune deficiency syndrome) A condition caused by a virus which weakens the body's immune responses. This results in reduced resistance to diseases such as pneumonia and can be fatal. The virus enters the bloodstream as a result of using contaminated needles, by transfusion of infected blood or blood-products, or via sexual contact with an infected person.

alcohol abuse The habitual excessive consumption of alcoholic drinks, which can lead to diseases of the **liver**, **nervous system** and **alimentary canal**.

algae A photosynthetic plant group including **microscopic** types such as *Spirogyra* and *Euglena*, and also **multicellular** types, e.g. seaweeds. Algae are widely distributed as marine and freshwater **plankton**, while some seaweeds are edible and others are a source of **agar**.

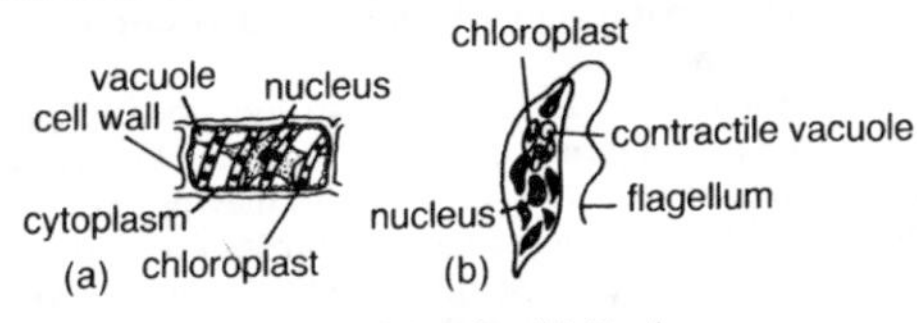

algae (a) *Spirogira*. (b) *Euglena*.

alimentary canal The digestive canal in animals. In humans it is a tube about nine metres in length running from **mouth** to **anus**. See **digestion**.

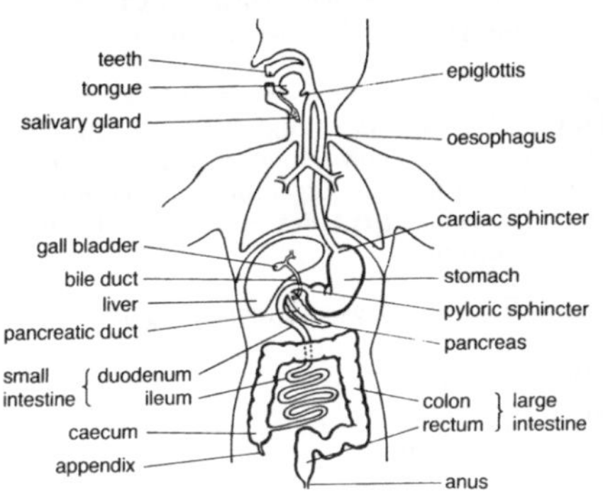

alimentary canal The human alimentary canal.

alkali An **aqueous solution** of a **base**, which releases hydroxyl ions in water and has a **pH** greater than 7, e.g. lime (calcium hydroxide), which is added to **soil** to *neutralize* excess **acid**.

alleles Each inherited feature in an individual is produced by a pair of **genes** or alleles, which are located on the **chromosomes**. The pair of genes may produce the same or differing effects. For example, the fruit fly *Drosophila* has a pair of alleles which control wing length. The combination of these dictates the possession of normal wings or vestigial wings. See **monohybrid inheritance**, **backcross**, **incomplete dominance**, **codominance**.

alveoli Air sacs in the mammalian **lungs** across which **gas exchange** occurs. See **gas exchange** (*mammals*).

amino acids **Organic compounds** which are the subunits of **proteins**. Altogether some 70 different amino acids are known, but only about 20–24 are actually found in living organisms, bonded together in chains known as **peptides** which are the basis of protein structure.

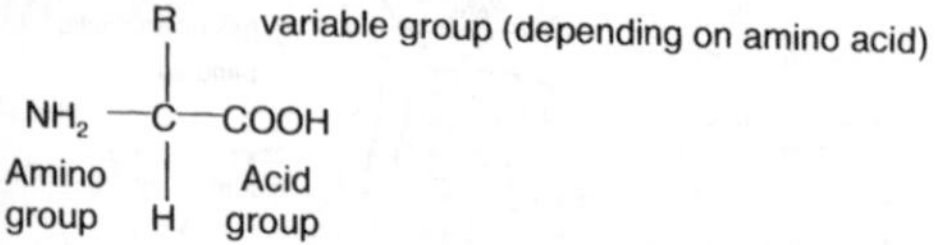

R variable group (depending on amino acid)

NH_2 — C — COOH

Amino group H Acid group

amino acid Structure.

amniocentesis During **pregnancy**, the removal of some fluid from the **amnion**. The fluid contains **cells**

from the **foetus**, and by studying their **chromosomes** abnormalities such as **Down's syndrome** or **inherited diseases** can be detected.

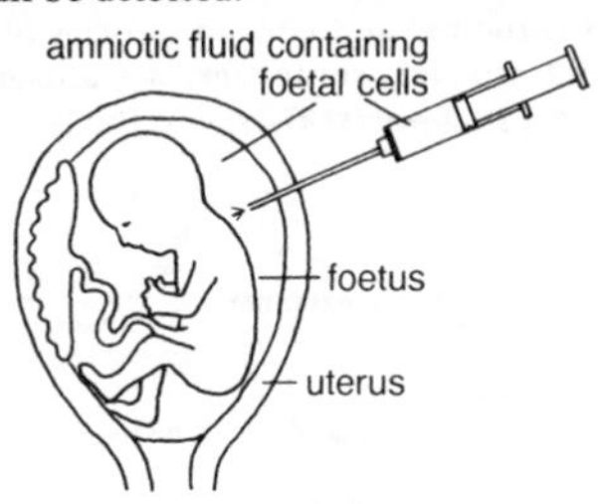

amniocentesis

amnion The fluid-filled sac surrounding and protecting the **embryos** of mammals, birds and reptiles. See **pregnancy**.

anabolism See **metabolism**.

anaerobe An organism which lives in the absence of oxygen. See **aerobe**, **respiration**.

annual A flowering plant which completes its **life history** from **germination** to death in one season.

antagonistic muscles Pairs of **muscles** which produce opposite movements, the contraction of one stimulating the relaxation of the other.
For example, at a **joint**, the contraction of the *flexor* (muscle which bends limb) stimulates the relaxation of

the *extensor* (muscle which straightens limb) so that bending occurs. When the joint straightens due to contraction of the extensor, this causes the flexor to relax. The longitudinal and circular muscles of the vertebrate **gut** and annelid worms also act antagonistically, the former causing **peristalsis**, and the latter movement.

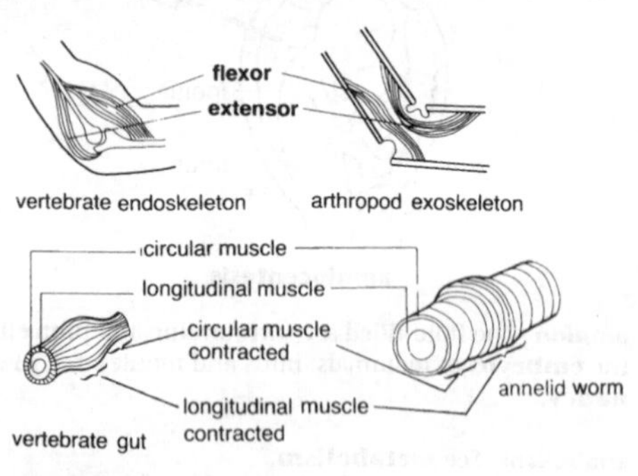

antagonistic muscles Pairs of muscles pulling in opposite directions perform various functions.

anterior Relating to parts of the body at or near the leading or head end of an animal. Compare **posterior**.

anther The upper part of a **stamen** in a **flower**, containing **pollen** grains.

antibiotics Substances formed by certain **bacteria** and **fungi** which inhibit the growth of other **microorganisms**. For example, **penicillin**, streptomycin.

antibiotic discs Sterile paper discs containing **antibiotic**, which are used to identify which antibiotic will be effective against infection with a particular **bacterium**. The discs are added to nutrient agar plates (see **agar**) which have been contaminated with bacteria taken from a patient.
In plate A, bacteria grow only around the **penicillin** disc, while the plate is clear around the *streptomycin* disc. This shows that not all antibiotics are effective against all **microorganisms**. This is confirmed by plate B, in which three of the antibiotics had no effect.

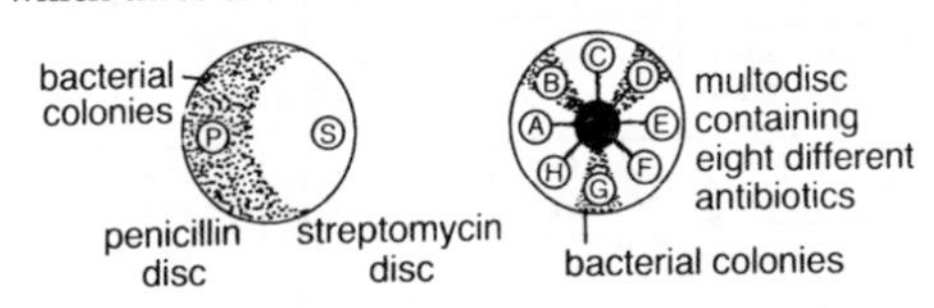

antibiotic discs Possible results after incubation. Certain antibiotics have destroyed the bacteria.

antibiotic resistance The ability of certain **microorganisms** to overcome the action of **antibiotics**: this results from **mutations** producing new strains which are no longer susceptible to the antibiotic.

antibodies **Proteins** produced by vertebrate **tissues** as

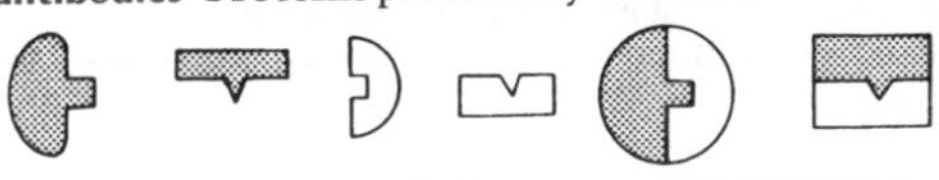

antibodies Antibodies react with antigens to make them harmless.

a reaction to **antigens**, i.e. materials foreign to the organism (for example, **microorganisms** such as bacteria and their **toxins**, or transplanted **organs** or tissues). See **agglutination**.

antidiuretic hormone (ADH) A **hormone** secreted in mammals by the **pituitary gland**, which stimulates water reabsorption by the **kidneys**, thus reducing water loss in the urine.

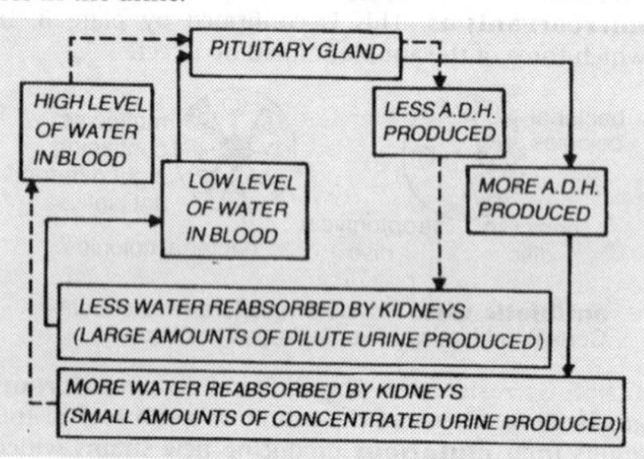

antidiuretic hormone Control of water loss.

antigen A substance, usually a **protein**, which stimulates the immune system to make an **antibody**. See **immunity**.

antiserum (*pl.* **antisera**) A substance containing large numbers of **antibodies** to a particular *antigen*. See **blood tests**.

anus The terminal opening of the **alimentary canal** in mammals, through which **faeces** are shed. The anus is opened and closed by a **sphincter muscle**.

aorta The largest **artery** in the **circulatory system** of mammals; it carries **blood** from the left **ventricle** of the **heart** to the rest of the body.

appendix A small sac found at the junction of the **ileum** and **caecum** of some mammals. In humans its function is unclear.

aqueous humour Clear watery fluid filling the front chamber of the vertebrate **eye** between the **cornea** and the **lens**.

artery A vessel which transports **blood** from the **heart** to the **tissues**. In mammals, arteries carry *oxygenated* blood, i.e. blood which carries oxygen in the haemoglobin to the body (for an exception to this rule, see **pulmonary vessels**) and divide into smaller vessels called *arterioles*. They have thick, elastic, muscular walls, in order to withstand the high pressure caused by the heartbeat.

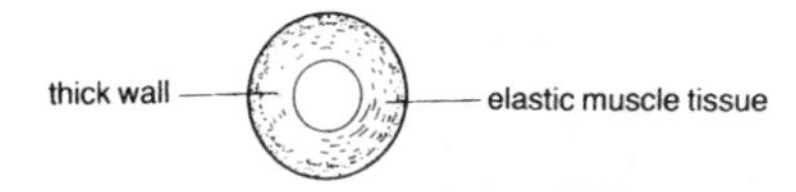

artery Section through an artery.

arterioles See **artery**.

artificial propagation The method by which plant growers make use of a plant's capacity for **asexual reproduction** and **regeneration** in order to produce new plants. Small pieces of stems, roots or leaves are taken and under the right conditions these will grow

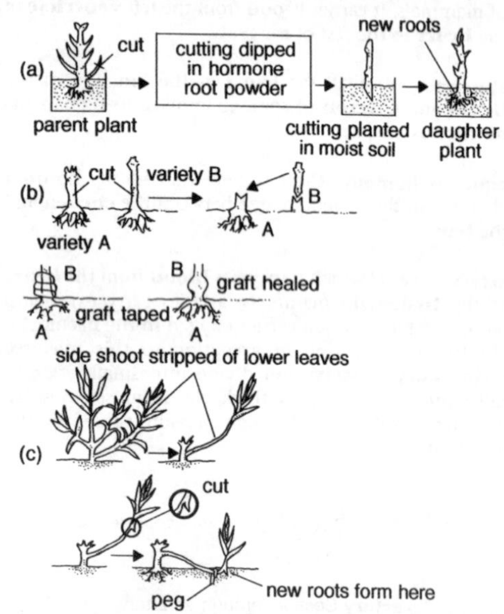

artificial propagation (a) Cutting. (b) Grafting. (c) Layering.

into new plants (**clones**). Three methods are shown here.

artificial selection The method by which animal and plant breeders attempt to improve stocks by selecting the males and females with desirable characteristics and allowing them to interbreed. By rigorous selection over many generations, improvements can be made in stock quality, e.g. in beef and milk production in cattle or yield and disease resistance in crops. See **natural selection**.

asexual reproduction **Reproduction** in which new organisms are formed from a single parent without **gamete** production. The offspring from asexual reproduction are genetically identical to each other and to the parent organism, and are referred to as **clones**. See **binary fission**, **budding**, **spore**, **vegetative reproduction**.

assimilation (of food) The process by which digested food particles are incorporated into the **protoplasm** of an organism. For example, in mammals glucose not required immediately to provide energy in tissue respiration is converted in **liver** and **muscle cells** to the storage **carbohydrate**, **glycogen**, which can be reconverted to glucose if the blood glucose level falls (see **insulin**). Surplus glucose not stored as glycogen is converted to **fat** and stored in fat storage cells beneath the **skin**, as a long-term energy store.
Fatty acids and glycerol (see **fats**) are reassembled into fat. Surplus fat is stored as outlined above.
Amino acids are synthesized into **proteins**. Surplus amino acids cannot be stored and are disposed of in the

liver by **deamination**.

atom The smallest complete particle of an **element** that can exist chemically. Each atom consists of a nucleus of **protons** and **neutrons** surrounded by moving **electrons**. See **molecule**.

ATP (adenosine triphosphate) A chemical compound which acts as a store and a source of **energy** within **cells**. ATP is formed from *adenosine diphosphate* (*ADP*) and a phosphate group using energy from **respiration** which can then be released for metabolic processes when ATP is broken down.

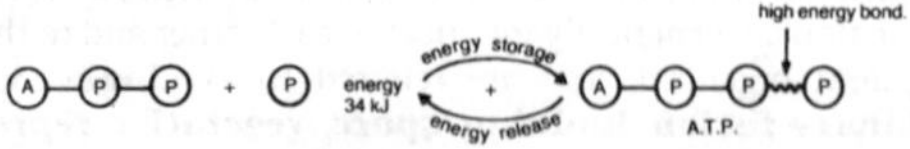

ATP ATP provides energy for metabolic processes.

atrium See **heart, heartbeat**.

auditory Relating to part of the body and functions connected with the **ear**.

auditory canal A tube in the mammalian outer ear leading from the **pinna** to the **tympanum**.

auditory nerve A **cranial** nerve in vertebrates conducting **nerve impulses** from the inner ear to the **brain**. See **ear**.

auricle See **heart, heartbeat**.

autoclave A pressure cooker used for the **sterilization** of materials e.g. **agar** before and after use in **microbiology** experiments. The materials to be sterilized are heated in the autoclave at 120 °C for 15 minutes to destroy any **bacteria** present.

autoradiograph A picture obtained when a photographic negative is exposed to living **tissue** into which radioactive material has been introduced in order to trace the route of substances through the tissue.

autotrophic (used of organisms) Able to synthesize complex **organic compounds** from simple non-living **inorganic compounds**. The major autotrophs are green plants, which use water and carbon dioxide to make food by **photosynthesis**. For this reason green plants are also called **food producers**. Compare **heterotrophic**.

auxins **Plant growth substances** which control many aspects of plant growth, for example, **tropisms**, by stimulating **cell division** and elongation.

axon See **neurones, synapse**.

backbone See **vertebral column**.

backcross A **genetic** cross in which a **heterozygous** organism is crossed with one of its **homozygous** parents. Thus two backcrosses are possible.
For example, in the fruit fly *Drosophila*, normal wings are **dominant** to vestigial wings and so heterozygous flies will have normal wings. The backcross with the

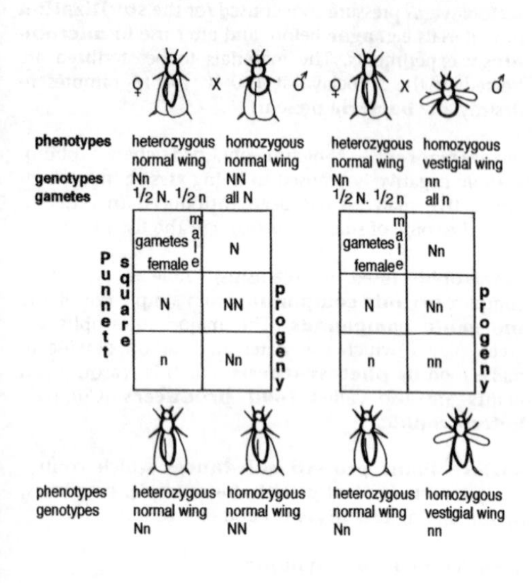

backcross The two backcrosses of *Drosophila*.

recessive homozygote is useful in distinguishing between organisms with the same **phenotype** but different **genotypes**. Examples are NN and Nn. Such a cross is called a **testcross**.

See **monohybrid inheritance**.

bacteria **Unicellular** organisms with a diameter of 1–2 microns. Some bacteria cause disease such as tetanus, but others are useful as, for example, sources of **antibiotics**.

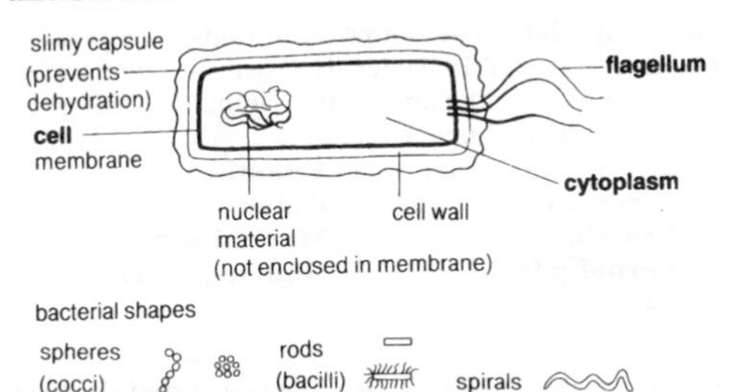

bacteria Structure of a generalized bacterium, with bacterial shapes.

Baermann funnel An apparatus used to isolate organisms living in soil water, e.g. **algae**, **protozoa**. The

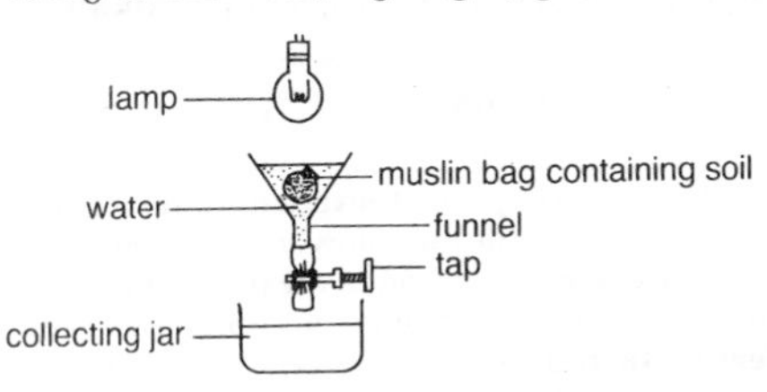

Baermann funnel

organisms move away from the strong light and high temperature of the lamp and collect near the tap, from where they can be released into the collecting jar.

balanced diet The correct nutritional components required for health, generally used in reference to human beings and domesticated animals. A balanced diet for humans should contain:

(a) Sufficient kilojoules of **energy**;
(b) **Protein**;
(c) **Carbohydrate**;
(d) **Fat**;
(e) **Vitamins**;
(f) **Water**;
(g) **Mineral salts**;
(h) **Roughage** (*fibre*)

basal metabolic rate (BMR) The rate of **metabolism** of a resting animal as measured by oxygen consumption. BMR is the minimum amount of **energy** needed to maintain life and varies with **species**, age and sex.

base A **compound**, usually a metallic oxide or hydroxide, which, if soluble in water, forms an **alkali**.

base pair See **genes**.

batch processing (in **biotechnology**) an industrial process in which the **enzymes** are dispersed throughout the **substrate**. At the end of the batch, the enzyme has to be separated from the product, and the **fermentation** vessel must be emptied and cleaned before the next batch. See **continuous flow processing**.

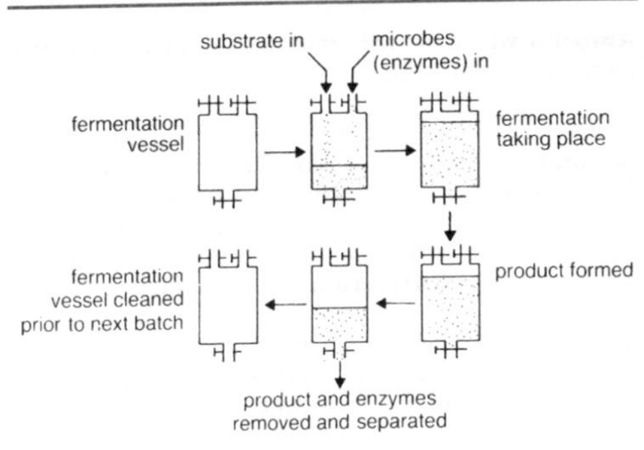

batch processing

bile A green alkaline fluid produced in the **liver** of mammals. Bile is stored in the **gall bladder** and is transported via the *bile duct* to the **duodenum** where it causes **fat** to be broken into minute droplets (*emulsification*) before **digestion**.

binary fission **Asexual reproduction** in **unicellular** organisms in which a single **cell** divides to produce two cells. The nucleus divides by **mitosis**. Binary fission is common in **bacteria** and **protozoa** such as

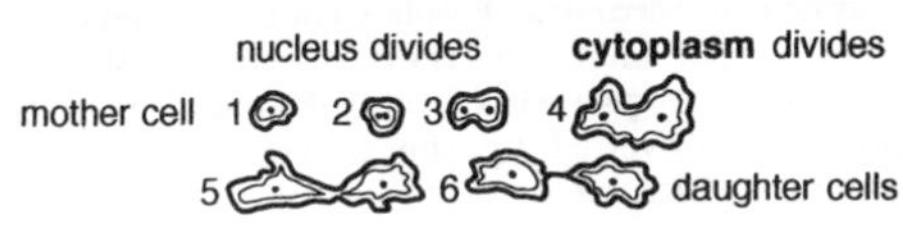

binary fission *Amoeba.*

Amoeba where a single mother cell divides into two identical daughter cells

binomial nomenclature The method of naming organisms devised by Carl Von Linne (Linnaeus) in the 18th century. Each organism has two Latin names, the first, with an initial capital, indicating the **genus**, and the second with a lower case first letter indicating **species**. (See **classification**.) For example:

Genus	Species	Common name
Canis	*familiaris*	domestic dog
Canis	*lupus*	American Wolf

biodegradable Able to be readily broken down by biological action. For example, biodegradable detergents in **sewage** can be digested by **bacteria**.

biological clock The mechanism thought to be responsible for animal **rhythmical behaviour** patterns associated with repeating natural cycles such as tides, **photoperiods** and seasons.

biological control The use of natural **predators**, **pathogens** and **parasites** to control **pests**.

biological detergents Powders containing **enzymes** obtained from **bacteria**. These enzymes break down stains caused by **proteins** in milk, **blood**, egg etc. The stains are converted by the enzymes into soluble substances which can be washed away

biomass The total mass of living matter in a **popula-**

tion. Biomass is usually expressed as living or dry weight and decreases at each level in a **food chain**. See **pyramid of numbers, pyramid of biomass**.

biosensor A device used for the rapid detection of chemicals in **blood** or **urine**. A biosensor uses **immobilization** to detect a particular substance by reacting specifically with the substance to give a product which is used to generate an electrical signal. An example of a biosensor is the *glucose oxidase electrode* which measures the amount of glucose in blood.

biosphere That part of the Earth which contains living organisms. The biosphere includes all the various **habitats** from the deepest oceans to the highest mountains. See **environment**.

biotechnology The use of living organisms in manufacturing processes. For example, **microorganisms** are used in ethanol production by **fermentation**, **brewing** with **yeast**, and **enzyme** production.

biotic factor See **environment**.

birth (in humans) The process by which a **foetus** leaves the **uterus** to live outside the mother as an individual being. The human baby is born as a result of muscular contractions of the uterus wall. The *amniotic fluid*, in which the baby has been floating, escapes, and the baby is pushed through the **cervix** and the **vagina** and thus leaves the mother's body.
The **umbilical cord** is cut, the **placenta** is expelled as the *afterbirth* and the baby must now use its own **lungs**

for **gas exchange**. See **fertilization**, **pregnancy**.

birth rate (of a **population**) the number of live births, measured in the human population as number of births in one year per 1000 of population. See **human population curve**.

bladder (urinary) A sac into which **urine** from the **kidneys** passes via the **ureters**. From the bladder, urine is discharged through the **urethra**. See **kidney**.

blind spot The blind spot in the **eye** can be demonstrated by a simple experiment, as follows.
Hold the book at arm's length. Close the left eye and concentrate on the cross with the right eye. Slowly bring the book closer until the drawing of the face seems to disappear. At this point the image of the face is falling on the blind spot.

blind spot

blood A fluid **tissue** found in many animals with the principal function of transporting substances from one part of the body to another. In mammals, blood consists of a watery solution called **plasma**, in which there are three types of cells: **platelets**, **red blood cells** and **white blood cells**.
The main functions of blood are:
(a) Transport of oxygen from **lungs** to **tissues**;
(b) Transport of toxic by-products to the **organs** of

excretion;

(c) Transport of **hormones** from **endocrine glands** to target organs;

(d) Transport of digested food from the **ileum** to the tissue;

(e) Prevention of infection by **blood clotting**, **phagocytosis** by white blood cells, and **antibody** production.

blood clotting The conversion of blood **plasma** into a clot, which occurs when blood **platelets** are exposed to air as a result of injury. The platelets produce an **enzyme** (*thrombin*) which causes the conversion of a soluble **plasma protein** (**fibrinogen**) into *fibrin*, which forms a meshwork of fibres and the resulting clot restricts blood loss and the entry of **microorganisms**.

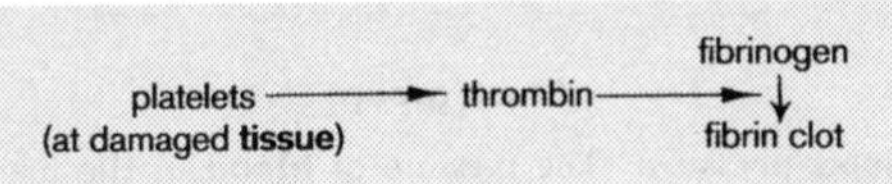

blood clotting The process of clot formation.

blood groups The classification of **blood** types based on the different antigens (see **antibodies**) present on the surface of **red blood cells**.

The human **population** is divided into four groups called A, B, AB and O. The capital letters stand for the type of antigen present in the red blood cells. The corresponding antibodies are carried in the **plasma**, and if a person has a particular antigen in his red cells, he cannot have the corresponding antibody since **agglutination** would occur. Thus

Group A contains antigen A and antibody anti-B.
Group B contains antigen B and antibody anti-A.
Group AB contains antigens A and B and no antibodies of either type.
Group O has no antigens and antibodies anti-A and anti-B.
See **blood transfusion**.

blood group	antigen on red blood cells	antibody in plasma
A	A	anti-B
B	B	anti-A
AB	A and B	neither
O	neither	anti-A and anti-B

blood groups

blood pressure The pressure of **blood** in the main **arteries** of mammals. In humans, blood pressure is normally about 120 mm mercury (Hg) at **systole** and about 80 mm Hg at **diastole**, but can vary with age, **exercise** etc. See **heartbeat**.

blood serum Fluid consisting of blood **plasma** with the **fibrinogen** removed.

blood tests The use of **agglutination** in order to determine a person's **blood group**. If one drop of blood from a sample is mixed with Anti-A **blood serum** and another drop with anti-B blood serum then:

Group A **red blood cells** will clump only in anti-A;
Group B red blood cells will clump only in anti-B;
Group AB red blood cells will clump in both **antisera**;
Group O red blood cells will clump in neither antiserum.
See **agglutination, blood groups, blood transfusions**.

blood group under test	antigens on cells	antibodies in plasma	added to anti-A test serum	added to anti-B test serum
A	A	anti-B		
B	B	anti-A		
AB	A and B	neither		
O	neither	anti-A anti-B		

blood tests Summary.

blood transfusion The transfer of **blood** from a healthy person (the *donor*) to another person who has lost a lot of blood, e.g. as a result of an injury. The **blood groups** of the donor and the patient must match or else the **antibodies** in the patient's **plasma** will act upon the **antigens** on the donor's **red blood cells** and cause the cells to clump together (**agglutination**).
See **blood groups**.

blood group	can donate blood to:	can receive blood from:
A	A and AB	A and O
B	B and AB	B and O
AB	AB	all groups
O	all groups	O

blood transfusion Table showing the pattern of blood acceptability.

blood vessels Tubes transporting **blood** around the bodies of many animals, which together with the **heart** make up the **circulatory system**. In vertebrates, the blood vessels consist of **arteries**, *arterioles*, **capillaries**, *venules* and **veins**.

BMR See **basal metabolic rate**.

bone Tissue in the vertebrate **skeleton** consisting of **collagen** (a **protein**) which gives **tensile strength**, and calcium phosphate which gives bone its hardness. Some bones have a hollow cavity containing bone marrow in which new **red blood cells** are produced.

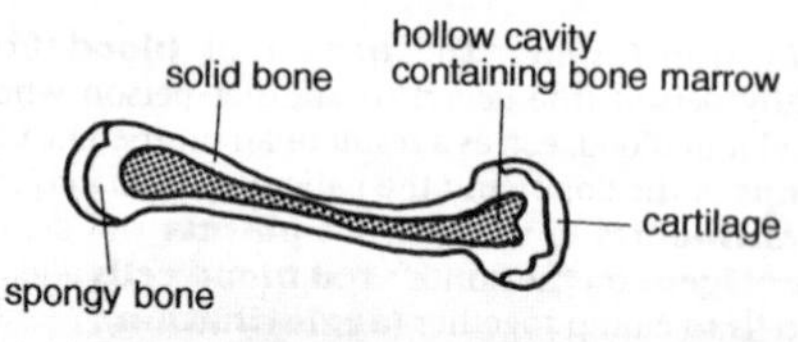

bone Section through bone.

Bowman's capsule A cup-shaped part of a **kidney** tubule or **nephron** in mammals.

brain The large mass of **neurones** in animals which has a centralized coordinating function. In vertebrates it is found at the **anterior** end of the body, protected by the **cranium**, and connected to the body via the **spinal cord** and *spinal nerves*, and directly by nerves called *cranial nerves*, e.g. **optic nerve**, **auditory nerve**.
The human brain contains millions of nerve cells which are continually receiving and sending out **nerve impulses**. The remarkable property of the brain is that it translates electrical impulses in such a way that stimuli from the **environment**, such as sound and light, are appreciated so that the recipient of the stimuli can respond and adapt to the environment in the most appropriate way. The brain also coordinates bodily activities to ensure efficient operation, and stores

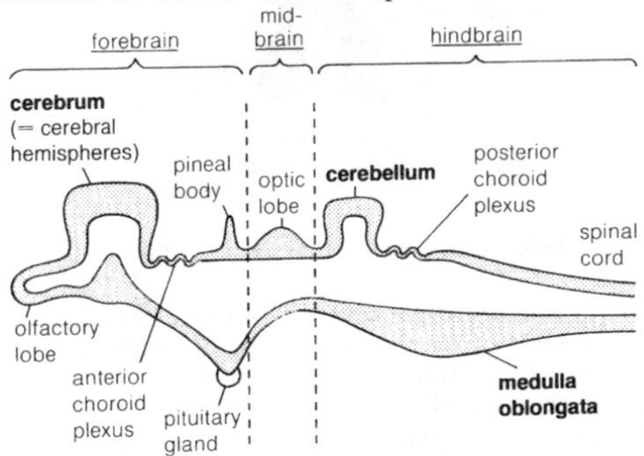

brain Diagram of the vertebrate brain.

information so that behaviour can be modified as the result of experience.

breastbone See **sternum**.

breathing (in mammals) The inhalation and exhalation of air for the purpose of **gas exchange**. In mammals the gas exchange surface is situated in the **lungs**. The exchange of air in the lungs (*ventilation*) is caused by changes in the volume of the **thorax**, brought about by the action of the **diaphragm** and **intercostal muscles**.

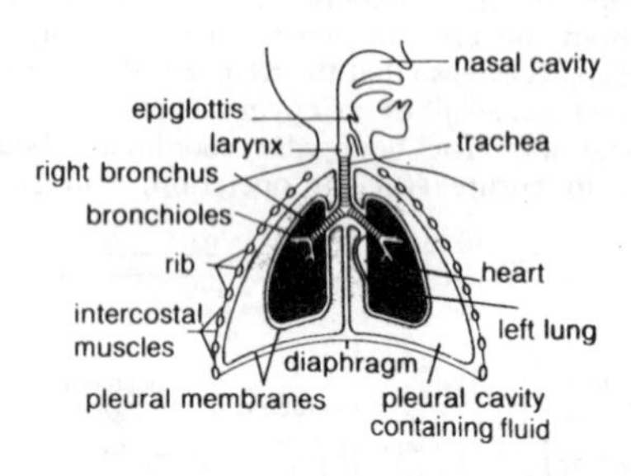

breathing Lungs and associated structures.

When the diaphragm contracts, it depresses, increasing the volume of the thorax, causing air to rush into the lungs. Relaxation of the diaphragm reduces the volume of the thorax, and causes exhalation of air. The action of the diaphragm is accompanied by the raising and lowering of the rib cage, which is necessary to accommodate the changes in lung volume. These rib cage movements are caused by contraction and relaxation of the intercostal muscles.

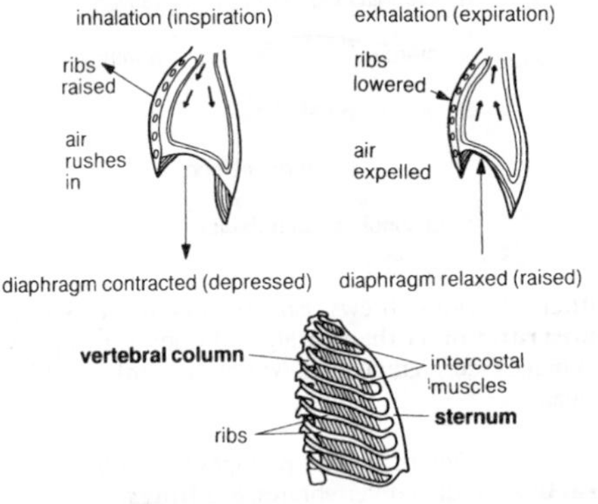

breathing Movements of the rib cage.

breathing rate The rate of **lung** ventilation. In humans, **breathing** movements are controlled by the **medulla oblongata** in the **brain**, which is sensitive to the carbon dioxide concentration of the **blood**. If the carbon dioxide concentration rises sharply as the result of increased **respiration**, for example, during exercise, the brain sends **nerve impulses** to the **diaphragm** and **intercostal muscles** which react by increasing the rate and depth of breathing. This accelerated breathing rate helps to expel the excess carbon dioxide and increases the supply of oxygen to respiring cells.

brewing The manufacture of beer by **fermentation**. The main stages are as follows.

1) Germinating barley grains convert

starch $\xrightarrow{\text{enzymes}}$ sugar (*malting*)

2) sugar + **yeast** + hops

↓ fermentation

alcohol + carbon dioxide
(beer)

Other alcoholic brews can be manufactured using **substrates** other than barley and hops, e.g. apple juice (to make cider), grape juice (wine), rice (*sake*) and honey (mead).

bronchus One of two air passages branching from the **trachea** in lunged vertebrates. See **lungs**.

budding **Asexual reproduction** in which a new organism develops as an outgrowth or *bud* from the parent, the offspring often becoming completely detached from the parent. Budding is common among coelenterates, e.g. *Hydra* and **unicellular fungi**, e.g. **yeast**. See diagram.

buffer A **solution** which maintains a constant **pH** even on the addition of an **acid** or an **alkali**.

bulb The **organ** of **vegetative reproduction** in flowering plants, consisting of a modified **shoot** whose

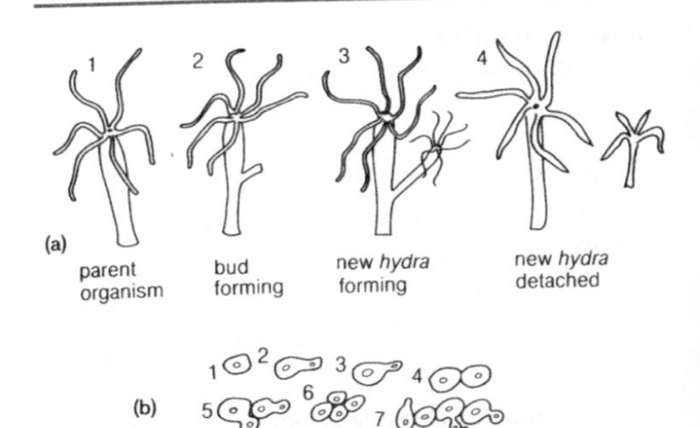

budding (a) *Hydra*. (b) Yeast.

short **stem** is enclosed by fleshy scale-like **leaves**. In the growing season, one or more buds within the bulb develop into new plants, using food stored in the bulb. Bulb-producing plants include the tulip, daffodil and onion.

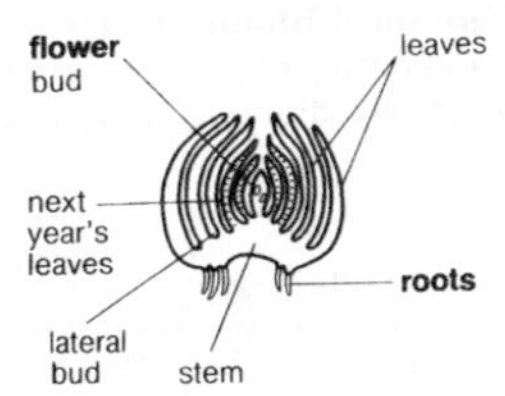

bulb Section through bulb.

caecum Part of the mammalian **gut** at the entry to the large **intestine**. In **herbivores** it is very important in

cellulose digestion. In humans it is much reduced in size and its function is uncertain.

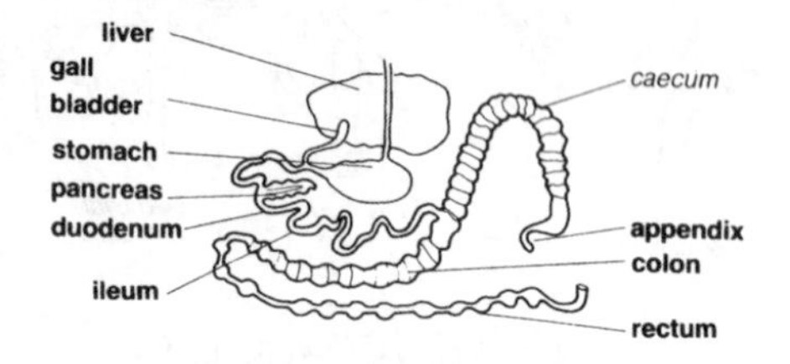

caecum The digestive system of herbivores.

canines Sharp-pointed tearing **teeth** near the front of the mouth used for killing prey, and ripping off pieces of food. Often reduced or missing in **herbivores**, present in **omnivores** and prominent in **carnivores**. See **dental formula, dentition**.

capillaries Very small **blood vessels** that branch off from *arterioles* and form a network in vertebrate **tissues**, the **blood** eventually draining into *venules* and then **veins**.

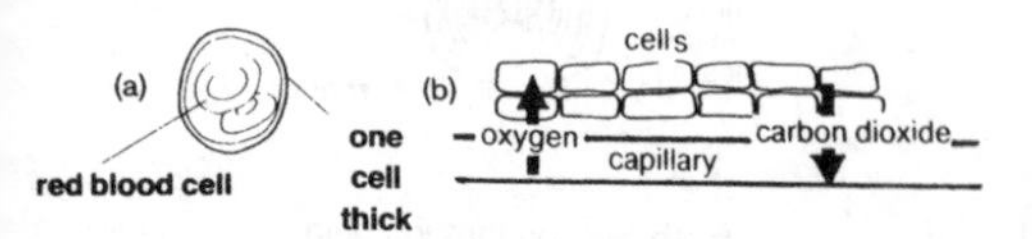

capillaries (a) section through capillary. (b) Capillary and tissues.

Capillary walls are only one **cell** thick, allowing diffusion of substances between the blood and the tissues via a liquid called *tissue fluid* (**lymph**).

carbohydrase Any **enzyme** which breaks down **carbohydrate**, by **hydrolysis**, into **disaccharides** and **monosaccharides**; examples are salivary amylase, maltase.

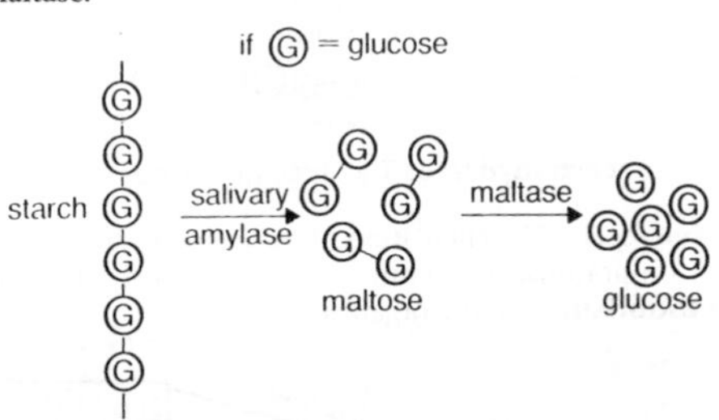

carbohydrase The action of salivary amylase and maltase.

carbohydrates Organic compounds containing the elements carbon (C), hydrogen (H) and oxygen (O) and with the general formula CH_2O. Carbohydrates are either individual **sugar** units or chains of sugar units bonded together. The three main carbohydrate types are **monosaccharides**, **disaccharides** and **polysaccharides**.

Importance of carbohydrates:

(a) Simple carbohydrates, particularly **glucose**, are the principal **energy** source within cells;

(b) Long-chain carbohydrates form some structural cell components, for example, **cellulose**, in plant cell walls, and also act as food reserves, for example, **glycogen** in animals and **starch** in plants.

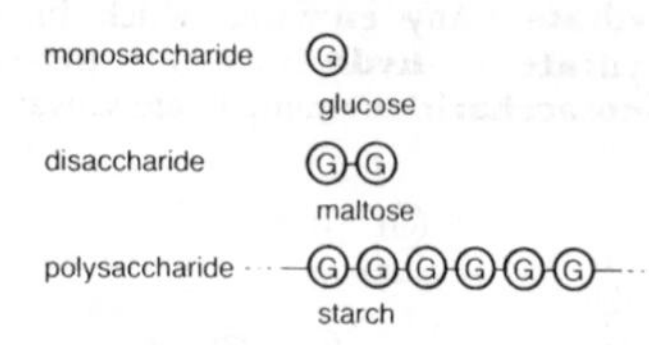

carbohydrates The three main types.

carbon cycle The circulation of the element carbon and its compounds, in nature, caused mainly by the **metabolism** of living organisms.

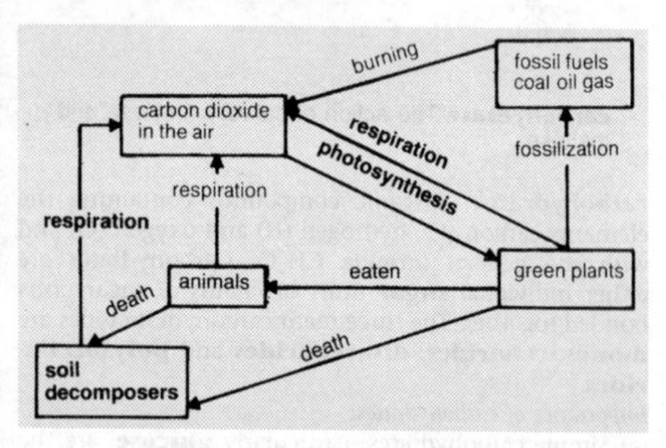

carbon cycle The main steps.

cardiac Relating to activities and parts of the body connected with the **heart** and its functions.

carnassials Shearing **teeth** used for cutting meat into chunks. These teeth are typical of **carnivores** and replace the premolars and molars found in **herbivores** and **omnivores**. See **dental formula, dentition**.

carnivore An animal that feeds on flesh. Carnivores include dogs, cats, etc. They have a **dentition** adapted for killing prey, shearing raw flesh, and cracking bones. The outstanding features of carnivore dentition are the large piercing **canine teeth**, and the shearing **carnassial** teeth. The lower jaw can usually only move up and down, forming an effective clamp on the prey. Carnivores typically have two sets of teeth during their lives.

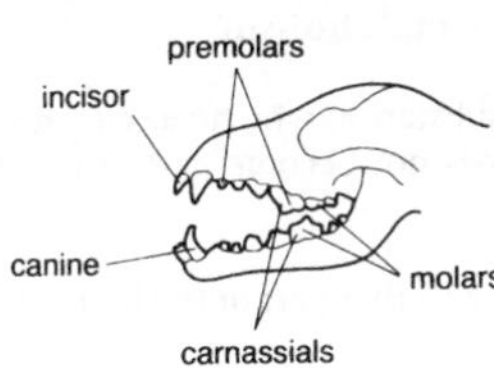

carnivore The skull and teeth of the dog.

carpel The female part of a **flower** containing an **ovary** in which there are varying numbers of **ovules** containing **embryo sacs**, within which are the female **gametes**.
See **fertilization**.

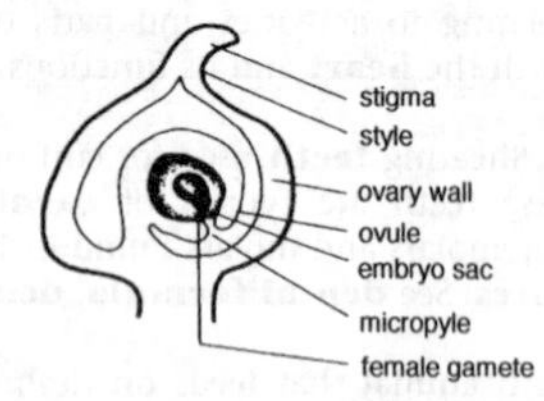

carpel The female reproductive part of a flower.

cartilage Supporting **tissue** found in vertebrates. In mammals there is cartilage in the **larynx**, **trachea**, **bronchi** and at the ends of **bones** at moveable **joints**, while in some fish, e.g. sharks, the entire skeleton is cartilage.

catabolism See **metabolism**.

catalyst A substance which accelerates a chemical reaction but does not become part of the end product. See **enzyme**.

caudal Relating to that part of the body of an animal at or near the tail.

cell A unit of **cytoplasm** governed by a single **nucleus** and surrounded by a **selectively permeable membrane**. Cells are the basic units of which most living things are made.
The **nucleus** contains the hereditary material, the chromosomes, and controls the cell's activities.
The **cytoplasm** is the liquid 'body' of the cell in which

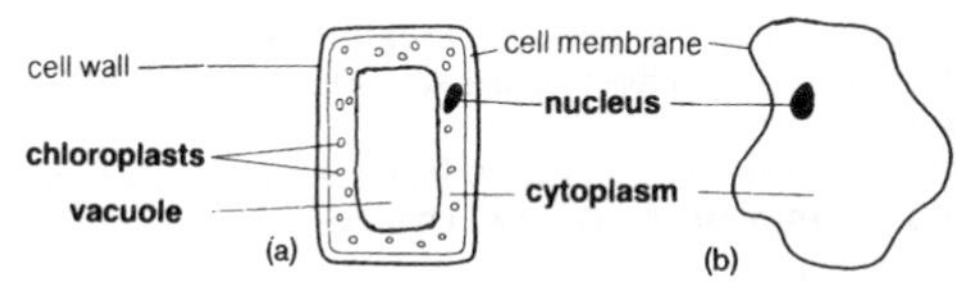

cell Structure of (a) a plant cell and (b) an animal cell.

the chemical reactions of life occur, for example, **respiration**.

The *cell membrane* controls the entry and exit of materials, allowing certain substances through, but preventing the passage of others. Such a membrane is described as a **selectively permeable membrane**.

The *cell wall* is found only in plants. It is made of **cellulose** and gives shape and rigidity to the cells.

The **chloroplasts** are structures within green plant cells where **photosynthesis** occurs.

The **vacuole** is filled with cell sap in plants. The sap, when in sufficient quantity, creates a pressure on the cytoplasm and cell wall and helps keep the cell firm and resilient. See **protoplasm**.

cell differentiation The process of change in **cells** during growth and development, whereby previously undifferentiated cells become specialized for a particular function as a result of structural changes.

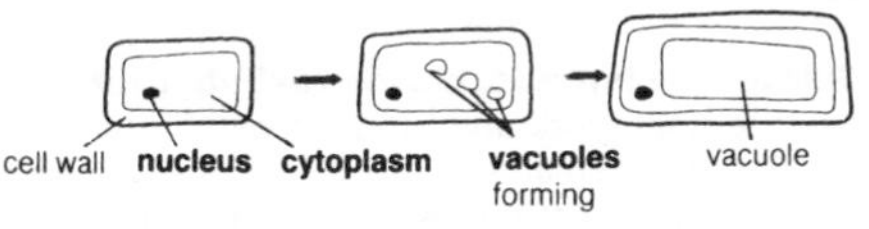

cell differentiation Elongation in plant cells.

For example in plant cells, after **cell division**, the daughter cells increase in size (*elongation*), by absorbing water.
After elongation, cell differentiation occurs as the result of **protoplasm** and cell wall changes.
For example:
(a) Some cells have their walls strengthened by additional **cellulose**, e.g. **cortex**, **epidermis**.
(b) Some cells have **lignin** deposited in their walls, e.g. **xylem**.
(c) Some cells develop extra **organelles**, for example, the **palisade mesophyll** cells of leaves develop a high number of **chloroplasts**.

cell division The division of the **cell** and its contents into two. The nucleus usually divides by **mitosis**, a process which gives the **nuclei** of the *daughter cells* exactly the same number of **chromosomes** as the *mother nucleus*. When a cell divides to form **gametes** (sex cells) the nucleus divides by **meiosis** which provides daughter nuclei with half the original number of chromosomes. In animals the **cytoplasm** divides by constriction into two. In plants a wall is laid down between the two halves.

cell membrane See **cell**, **selectively permeable membrane**.

cellulose The **polysaccharide carbohydrate** which forms the framework and gives strength to plant **cell** walls. Cellulose remains undigested in the human **gut** but has an important role as **roughage**. In mammalian **herbivores**, in the **caecum** and **appendix**, **bacterial**

cellulase
cellulose
glucose

cellulose The breakdown of cellulose into glucose by the action of cellulase.

populations produce an **enzyme** called *cellulase* which digests cellulose. See **digestion**.

cell wall See **cell**.

central nervous system (CNS) That part of the vertebrate **nervous system** which has the highest concentration of **neurone** bodies and **synapses**, i.e. the **brain** and **spinal cord**.

cerebellum The region of the vertebrate **brain** which in mammals controls balance and muscular coordination, allowing precise controlled movements in activities such as walking and running.

cerebrum or **cerebral hemispheres** The region of vertebrate **brain** which in mammals makes up the largest part of the brain. In humans the cerebrum consists of

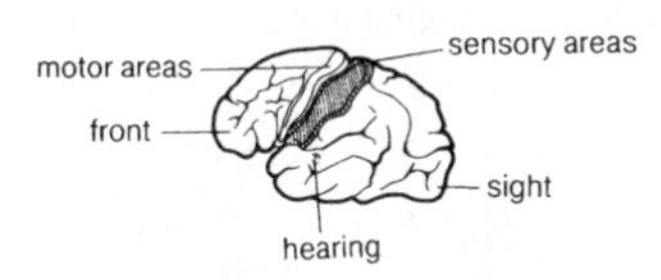

cerebrum Left cerebral hemisphere of human brain, showing localization of functions.

right and left hemispheres, the outer part made up of **neurone cell bodies** (*grey matter*), the inner part consists of nerve fibres (*white matter*). The human cerebrum is responsible for the higher mental skills such as memory, thought, reasoning and intelligence. The cerebrum also contains localized areas concerned with specific functions. Areas receiving **nerve impulses** from **receptors** are called *sensory areas*, while those sending out impulses to **effectors** are called *motor areas*.

cervical Describes parts of the body and functions related to (a) the neck, (b) the **cervix**.

cervix The posterior region of the mammalian **uterus**, leading into the **vagina**. See **fertilization**.

chemoreceptor A **receptor** which is stimulated by chemical substances. Examples are **smell** and **taste** receptors.

chemotropism **Tropism** relative to chemical substances. The growth of **pollen** tubes towards the **ovary** is an example of positive chemotropism. See **fertilization**.

chlorophyll Green pigment found in the **chloroplasts**

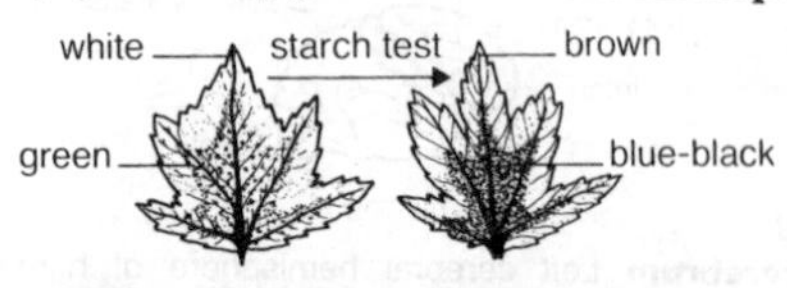

chlorophyll The variegated-leaf test.

of plant cells which can absorb the light **energy** required for **photosynthesis**. The importance of chlorophyll can be shown by the *variegated leaf test* for **starch**. Only those parts of the leaf which were previously green (i.e. which contained chloroplasts) give a positive starch test, showing that chlorophyll is necessary for photosynthesis.

chloroplasts Structures in the **cytoplasm** of green plant **cells**, in which **photosynthesis** occurs. See **palisade mesophyll**. Chloroplasts contain the green pigment **chlorophyll**.

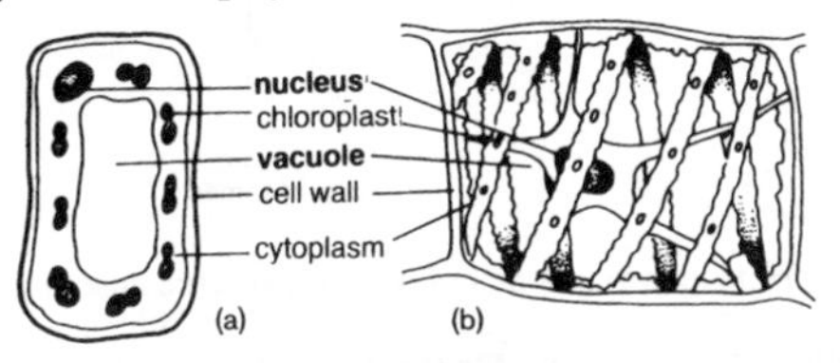

chloroplasts The chloroplasts of (a) a palisade mesophyll cell and (b) the alga *Spirogyra*.

choroid A layer of **cells** outside the **retina** of the vertebrate **eye**.

chromosomes The hereditary material within the **nucleus** of **cells**, which links one generation with the next. Each **species** has characteristic numbers and types of chromosomes.

For example, in humans, the chromosome number is 46. When a nucleus divides by **mitosis**, this **diploid** number of chromosomes is maintained in the new nuclei formed. **Haploid** nuclei contain half the diploid number of chromosomes, and are made when the nucleus divides by **meiosis**. Two haploid **gametes** join to form a diploid **zygote**.

Chromosomes control cellular activity. They consist of sub-units called **genes** which contain coded information in the form of the chemical compound **DNA**. In diploid cells, chromosomes occur in similar pairs known as homologous pairs. Thus a human diploid cell contains 23 pairs of **homologous chromosomes**.

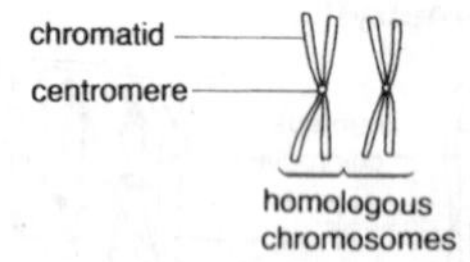

chromosomes A pair of homologous chromosomes.

cilia Microscopic motile threads projecting from certain **cell** surfaces which stroke rhythmically together like oars. Cilia occur in certain vertebrate **epithelia** where they cause movement of particles in the **trachea**, **oviduct**, **uterus**, etc. In some **protozoa** e.g. *Paramecium*, cilia cause movement of the whole organism. Compare **flagellum**.

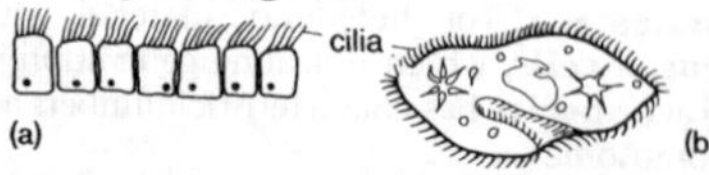

cilia The cilia of (a) epithelial cells and (b) *Paramecium*.

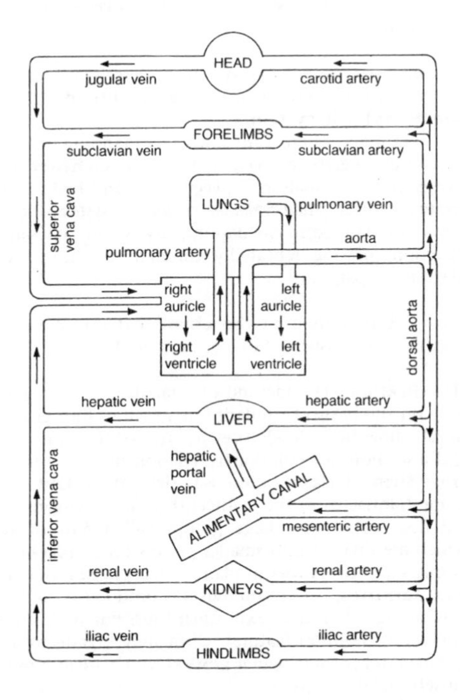

circulatory system The double circulatory system of mammals.

ciliary muscle **Tissue** in the vertebrate **eye** responsible for **accommodation**.

circulatory system Any system of vessels in animals through which fluids circulate, e.g. the **blood** circulation, **lymphatic system**.
In mammals there are two overlapping blood circulations, i.e. there is a circulation between **heart** and **lungs** and a circulation between heart and body. This arrangement is called a *double circulatory system*. Blood flows through both circulations, always in the same direction, passing repeatedly through the heart. See diagram on page 45.

class A unit used in the **classification** of living organisms, consisting of one or more **orders**.

classification The method of arranging living organisms on the basis of similarity of structure into groups which show how closely they are related to each other and also indicate evolutionary relationships. The modern system of classification was devised by Carl von Linne (Linnaeus) in the eighteenth century. Organisms are first sorted into large groups called **kingdoms** which are divided into smaller groups called **phyla** in animals and **divisions** in plants, then **classes**, **orders** and **families**, each subdivision producing subsets containing fewer and fewer organisms, but with more and more common features. Ultimately organisms are grouped in genera (singular **genus**) which are groups of closely related **species**. It is not uncommon for scientists to disagree as to how to classify certain organisms. See table.

	Human	Dog	Oak	Meadow buttercup
Kingdom	Animal	Animal	Plant	Plant
Phylum/Order	Chordata	Chordata	Spermatophyta	Spermatophyta
Class	Mammalia	Mammalia	Angiospermae	Angiospermae
Order	Primates	Carnivora	Fagales	Ranales
Family	Hominidae	Canidae	Fagaceae	Ranunculaceae
Genus	*Homo*	*Canis*	*Quercus*	*Ranunculus*
Species	*sapiens*	*familiaris*	*robur*	*acris*

classification How four organisms are classified.

clavicle or **collarbone** The ventral **bone** of the shoulder-girdle of many vertebrates articulating with the **scapula** and **sternum**. See **endoskeleton**.

cloaca The posterior region of the **alimentary canal** in most vertebrates (but excluding mammals) into which the terminal part of the **intestine** and the **kidney** and reproductive ducts open.

clone A group of organisms which are genetically identical to each other, having been produced by **asexual reproduction**.

cloning methods See **artificial propagation**.

cochlea A spiral structure in the mammalian inner **ear** containing an area called the *organ of Corti* in which are located **neurones** which are sensitive to sound vibrations.

codominance A situation in which both **alleles** are

phenotype (blood group)	genotype
A	I^AI^A or I^AI^O
B	I^BI^B or I^BI^O
AB	I^AI^B
O	I^OI^O

codominance The phenotypes and genotypes of the four blood groups using the symbol I to represent the alleles.

expressed equally in the **phenotype** of a **heterozygote**.
The inheritance of human **blood groups** includes an example of codominance. There are four blood group phenotypes. A, B, AB and O. The **genes** for groups A and B are codominant and both are completely dominant to the gene for group O. Thus if a person inherits genes for group A and group B, half his **red blood cells** will carry antigen A and half antigen B (see **antibodies**).

CNS See **central nervous system**.

cold blooded See **ectotherm**.

collagen Fibrous **protein**, which is the principal component of vertebrate **connective tissue**, and an important skeletal substance in higher animals, conferring **tensile strength** to **bones**, **tendons** and **ligaments**.

collarbone See **clavicle**.

colon A region of the **large intestine** in mammals between the **caecum** and **rectum**, which receives undigested food from the **ileum**. In the colon, much of the water is absorbed from the undigested food, and the semi-solid remains (**faeces**) are passed into the **rectum**. See **digestion**.

commensalism A symbiotic relationship in which one of the organisms benefits, while the other neither suffers nor benefits. For example, a marine worm lives in a shell with a crab, sharing the crab's food, but giving nothing in return. See **symbiosis**.

community The **population** of different **species** living in a particular **habitat** and interacting with each other. For example, a rockpool habitat may have a community made up of crabs, worms, sponges, seaweeds, etc. See **niche**.

companion cells Cells in flowering plants, that are associated with **phloem** *sieve tubes*, and are believed to contribute to the transport function of phloem (**translocation**).

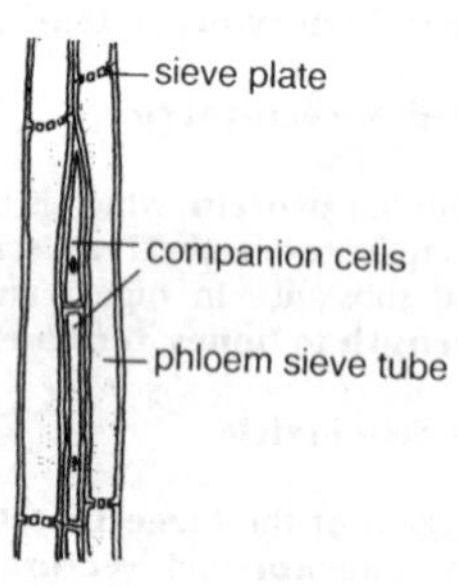

companion cell

compensation point (of green plants) The light intensity at which the rate of carbon dioxide uptake (**photosynthesis**) is exactly equal to the rate of carbon dioxide production (**respiration**). In a single day there are two compensation points when the rate of photosynthesis (**carbohydrate** gain) is exactly balanced by the rate of respiration (carbohydrate loss).

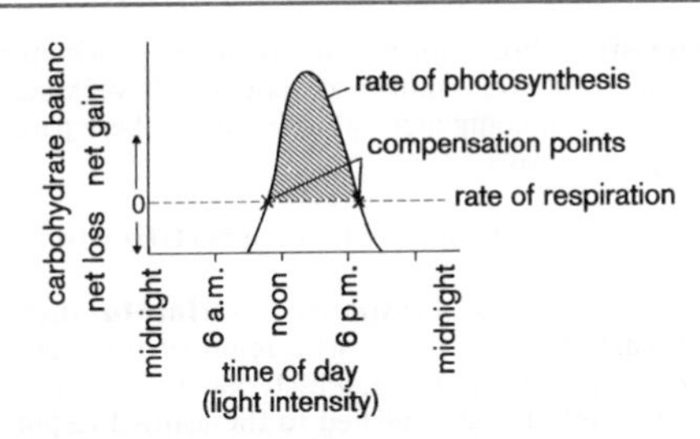

compensation point The two points at which the rates of photosynthesis and of respiration are equal.

competition The interaction among organisms of the same species (intraspecific competition) or organisms of different species (interspecific competition) seeking a common resource such as food or light which is in limited supply in the area occupied by the **community**. Competition often results in the elimination of one organism by another, or even in the elimination of one species as happens when two species of *Paramecium* compete for food.

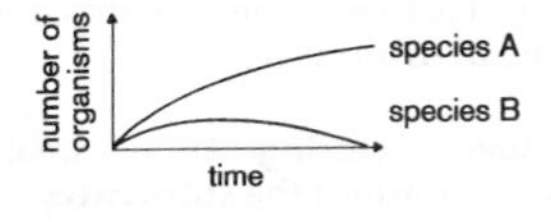

competition

compound A chemical formed by the combination of **elements** with the component **atoms** occurring in fixed proportions. The basic unit of a compound is the

molecule whose formation requires a chemical reaction. Mixtures, unlike compounds, have variable proportions of component atoms and can be separated by physical means.

conception An alternative term for **fertilization**.

conditioned reflex A **response** to a **stimulus** that has been learned by an animal as a result of the repeated association of the stimulus, which may be neutral, to a particular effect that is related to the learned response. For example, a rat may learn to press a lever when hungry as a result of learning to associate the lever's movement with the delivery of food. See **sensitivity**.

cone **1.** A reproductive structure of *gymnosperms*, e.g. pines.
2. A light-sensitive **neurone** in the **retina** of most vertebrate **eyes**. Sensitive in bright light, cones can detect colour.

connective tissue Supporting and packing **tissue** in vertebrates, consisting mainly of **collagen** fibres, in which are embedded more complex structures, such as **blood vessels**, **neurones** etc.

continuous flow processing An industrial process in **biotechnology**, in which the **substrates** flow continuously into the **fermentation** vessel and the product flows out continuously. This is made possible by using immobilized **enzymes** which can be used over and over again (see **immobilization**).
Continuous flow processing is more efficient than

batch processing since the enzymes can be used repeatedly, the product does not have to be separated from the enzymes, and no time-consuming turn-around is involved.
See **batch processing**, **immobilization**.

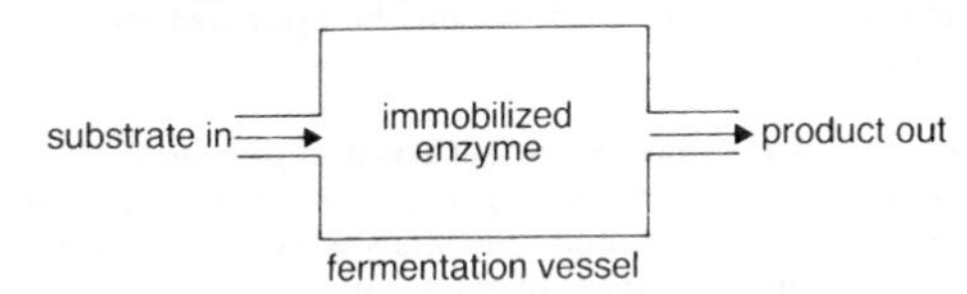

continuous flow processing

continuous variation See **variation**.

contractile vacuole A small sac in the **cytoplasm** of freshwater *Protista*, the function of which is **osmoregulation**, i.e., in response to water-entry by **osmosis**, the vacuole expands as it fills with water, and then contracts, discharging its contents out of the **cell**.

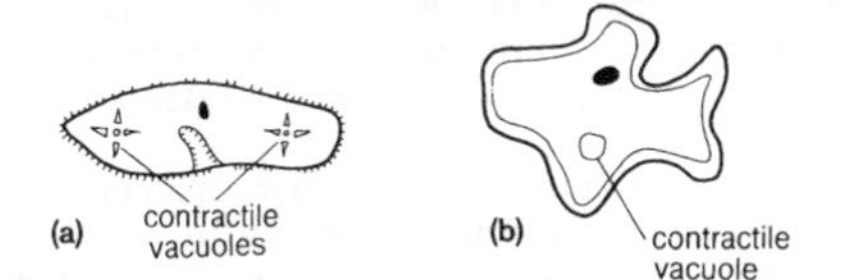

contractile vacuole The contractile vacuoles of (a) *Paramecium* and (b) *Amoeba*.

control experiment A test set up in a scientific investigation in which the factor being investigated is kept constant, so that the result of another test in which this factor is varied can be compared. See **scientific method**.

copulation The coupling of male and female animals for the purpose of **fertilization**. In humans the **penis** is inserted in the **vagina** and the **spermatazoa** are released.

corm An **organ** of **vegetative reproduction** in flowering plants consisting of an underground **stem** containing a food store and buds which develop into new plants. Examples of corms include those of the crocus and gladiolus.

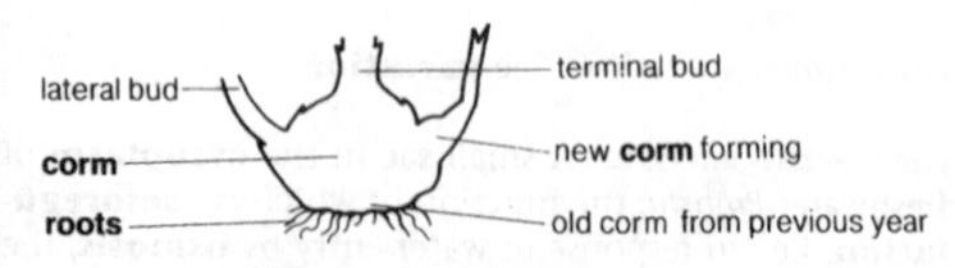

corm Section through a crocus corm.

cornea Transparent **tissue** at the front surface of the vertebrate **eye**, continuous with the **sclerotic** and involved in focusing the image on the **retina**.

cortex **1.** (in animals) The outer layer of an organ such as the mammalian **kidney**. See **medulla**.
2. (in plants) The layer of **cells** between the **epidermis** and the **vascular bundles**. Cortex cells are packing and supporting tissue, and in some cases, may store food. See **leaf, root, stem.**

cotyledon An embryonic **leaf** within a **seed** which

supplies food during **germination**, and in some plants is brought above the soil to carry out **photosynthesis** for a time before withering. Flowering plants with one cotyledon are called **monocotyledons**, and those with two are called **dicotyledons**.

courtship behaviour A type of animal behaviour which establishes contact between the sexes and thus increases the chance of successful breeding. Examples are bird song and display.

cranial Relating to activities and parts of the body connected with the **brain** and **cranium**.

cranium The **bones** of the vertebrate skull which enclose and protect the **brain**. See **endoskeleton**.

crop rotation The practice of growing a different crop in the same area in successive years in order to prevent **soil depletion**. Since different plants have different **mineral salt** requirements, changing the crop annually prevents depletion of one particular mineral salt. Another benefit is that since different plants have different **root** lengths, they absorb mineral salts from different **soil** depths. *Leguminous plants* such as peas, beans or clover are often included in rotations because of the **nitrogen fixation** within their **root nodules**. A typical crop rotation might be wheat/turnips/barley/clover/wheat, etc.

crossing over See **meiosis**.

cuticle A noncellular layer secreted by the **epidermis**

of above-ground plant structures, and by many invertebrates. Plant cuticles reduce water loss by **transpiration**, while invertebrate cuticles afford protection against mechanical damage and may also retain or repel water.

cytoplasm That part of the **protoplasm** of a **cell** bounded by the **cell membrane** but excluding the **nucleus.**

deamination Removal of the *amino* (-NH_2) group from surplus **amino acids**. In mammals, this occurs in the **liver**, the amino group automatically changing to the toxic compound *ammonia* (NH_3) which is then converted to **urea** and excreted. The remaining carbon-containing group is converted to useful **carbohydrate**.

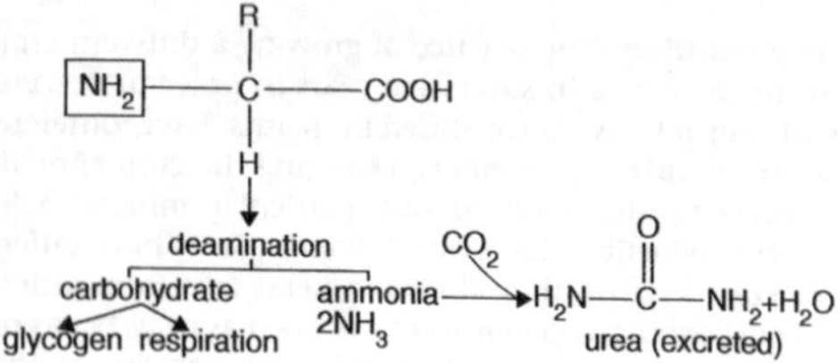

R
NH_2—C—COOH
H
deamination
carbohydrate
glycogen respiration
ammonia $2NH_3$ → CO_2 → H_2N—C(=O)—NH_2+H_2O
urea (excreted)

deamination The results of deamination in mammalian liver.

death rate (of a **population**) The number of deaths, measured in the human population as the number of deaths in one year per 1000 of population. See **human population curve.**

decay The process by which **decomposers** use the organic matter of dead organisms as a source of energy. See **carbon cycle, nitrogen cycle**.

deciduous teeth See **milk teeth**.

decomposers **Heterotrophic** organisms which cause the breakdown of dead animals and plants, and by so doing release their constituent compounds which can be used by other organisms. **Soil** decomposers include **bacteria**, earthworms, etc. See **carbon cycle, nitrogen cycle**.

denaturation Changes occurring in the structure and functioning of **proteins** (including **enzymes**) when subjected to extremes of temperature or **pH**.

dendrite or **dendron** See **neurones, synapse**.

denitrification The conversion, by **soil** *denitrifying bacteria*, of nitrates into nitrogen which can re-enter the atmosphere. See **nitrogen cycle**.

dental formula A formula describing the **dentition** of a mammal and expressed by writing the number of **teeth** in the upper jaw of one side of the mouth over the number of teeth in the lower jaw on one side. See diagram overleaf. The dental formula relates to an adult mammal with the full number of teeth. The total number of teeth is found by doubling the dental formula. See **carnivore, herbivore**.

dentition The numbers and types of **teeth** in a mam-

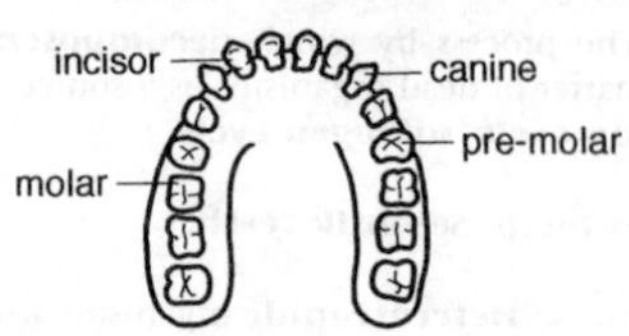

dental formula Teeth in human upper jaw. Dental formula: incisor $^2/_2$ canine $^1/_1$ premolar $^2/_2$ molar $^3/_3$ total number of teeth = 2 x dental formula = 2 x 16 = 32.

mal, described by a **dental formula**. Dentition reflects an animal's diet, i.e. an animal has the type of teeth best suited to deal with the type of food on which it feeds. See **carnivore, herbivore, omnivore.**

deoxygenated blood See **vein.**

deoxyribonucleic acid See **DNA.**

dialysis See **kidney machine.**

diaphragm A dome-shaped **muscle** separating the **thorax** and **abdomen** in mammals. Contraction and relaxation of the diaphragm is important in **lung** ventilation. See **breathing.**

diastema A toothless gap in the mouth of many **herbivores**, allowing the tongue to manipulate food more easily.

diastole See **heartbeat.**

dicotyledons One of the two subsets of flowering

plants, the other being **monocotyledons**. The characteristics of dicotyledons are
(a) two **cotyledons** in the **seed**;
(b) network of branching **veins** in **leaves**;
(c) broad leaves;
(d) ring of **vascular bundles** in **stem**;
(e) flower parts in fours or fives or multiples of these numbers.
Examples: hardwood trees, fruit trees, herbaceous plants.

diffusion The movement of particles from a region of high concentration to a region of lower concentration until they are evenly distributed. Diffusion occurs when two different particle concentrations are adjacent.

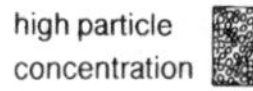

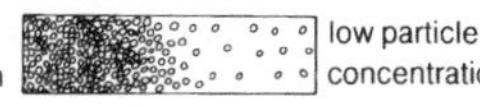

direction of particle movement

diffusion

The difference in concentration which causes diffusion is called a *concentration gradient*. The greater the concentration gradient, the greater is the rate of diffusion. If no concentration gradient exists, diffusion does not occur, and the situation is described as *equilibrium*.
Diffusion is the method by which many substances enter and leave living organisms, and are transported within and between **cells**. Examples are (a) uptake of water by plants from **soil**, (b) **gas exchange** between plants and the atmosphere, and (c) gas exchange between **blood** and respiring cells.
Where diffusion is too slow for a particular function, substances can be transported more rapidly by **active**

transport. For a special case of diffusion see **osmosis**.

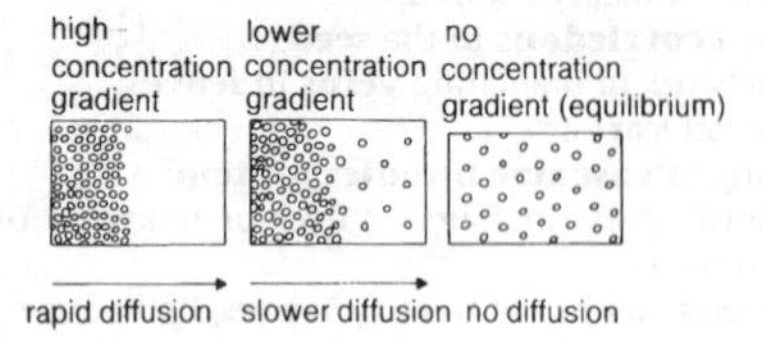

diffusion

digestion The breakdown by **enzyme** action of large insoluble food particles into small soluble particles, prior to **absorption** and **assimilation**. In many animals, including mammals, digestion and absorption occur in the **alimentary canal**. See table.

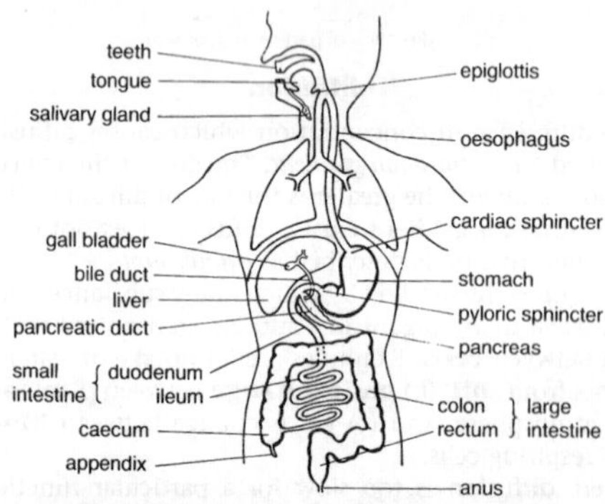

digestion The human alimentary canal.

Location	Glands	Enzyme	Substrate	Product
Mouth	Salivary	Amylase	Starch	Maltose
Stomach	Gastric	Pepsin	Protein	Peptides
		Rennin	Milk protein	Coagulated milk
Duodenum	Pancreas	Amylase	Starch	Maltose
		Lipase	Fats	Fatty acids+glycerol
		Trypsin	Protein+ peptides	Amino acids
Ileum		Lactase	Lactose	Glucose+galactose
		Lipase	Fats	Fatty acids+glycerol
		Maltase	Maltose	Glucose+fructose
		Peptidase	Peptides	Amino acids

digestion Digestive enzymes in humans.

diploid (used of a **nucleus, cell** or organism) Having the full complement of **chromosomes**, these occurring in pairs of **homologous chromosomes**. All animal cells, except **gametes**, are diploid. Gametes contain half the diploid number (**haploid**) as the result of **meiosis**. See **fertilization, mitosis.**

disaccharides Double **sugar carbohydrates** consisting of two **monosaccharides** linked together by bonds. For example, **maltose** is two **glucose** units joined together, while *sucrose* is one glucose unit linked with one fructose unit.

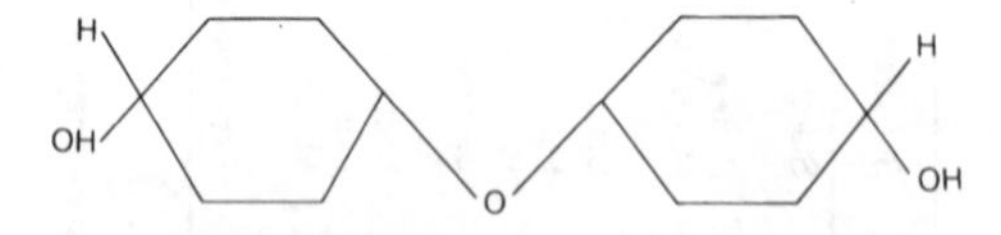

disaccharides Chemical structure.

disinfectant A chemical used to kill **microbes** outside the body, for example on surfaces such as floors.

division A unit used in the **classification** of plants. It is the equivalent of the term **phylum** used in the classification of animals. Divisions and phyla consist of one or more **classes**.

DNA (deoxyribonucleic acid) **Nucleic acid** which is the major constituent of **genes** and hence **chromosomes**. DNA consists of a double polynucleotide chain twisted into a *helix* (spiral) the two chains being held together by bonds between nitrogen *base pairs*.

These nitrogen bases can only link as complementary pairs: *thymine* with *adenine* and *guanine* with *cytosine*. The numbers and sequence of base pairs in the DNA polynucleotide chain represent coded information (the **genetic code**) which acts as a blueprint for the transfer of hereditary information from generation to generation.

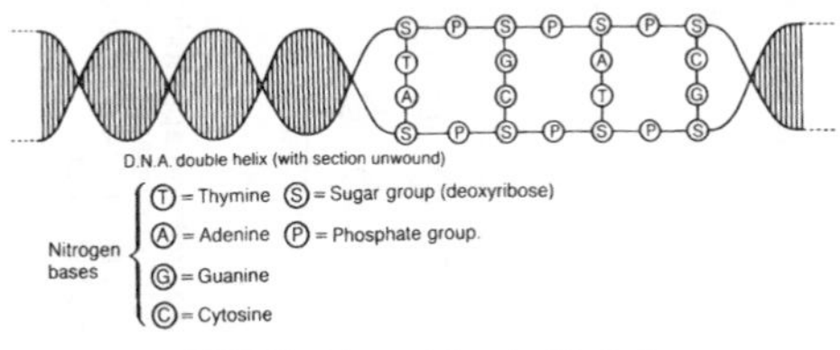

DNA Structure of the double helix.

DNA replication The formation of two new molecules of **DNA** during **mitosis**, each of which has exactly the same sequence of bases as the parent molecule. The process requires the four *nucleotides* (see **nucleic acids**), the appropriate **enzymes** and **ATP** for **energy**.

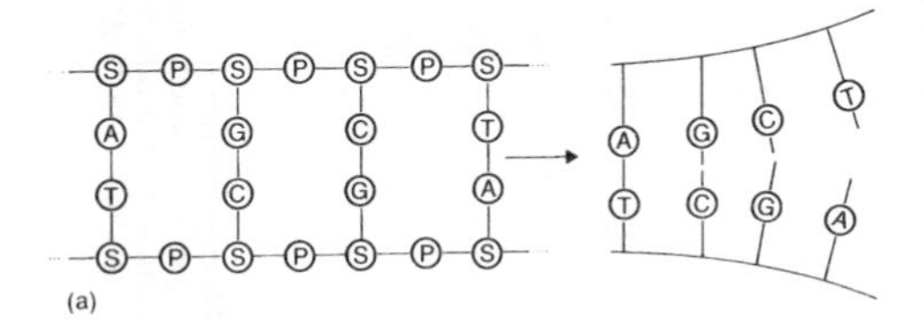

DNA replication (a) the DNA double helix unwinds and unzips.

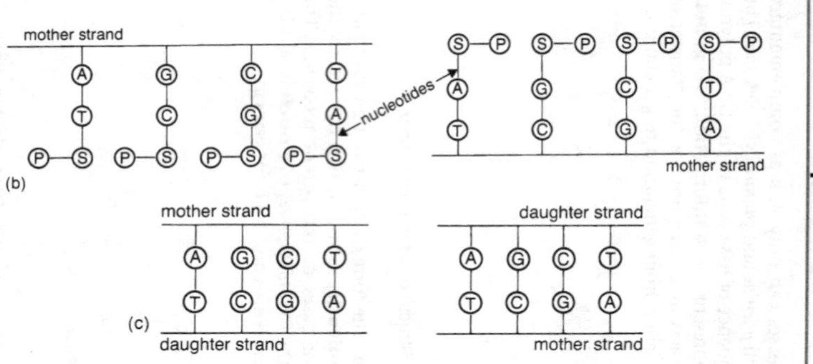

DNA replication (b) Each single chain then links with the appropriate nucleotide raw materials which are available in the cell. (c) After attachment of the bases, enzymes link the adjacent S and P molecules into two DNA double chains which are identical to each other and to the original 'mother' chain.

dominant (used of one of a pair of **alleles**) Always expressed in a **phenotype**, the other allele being described as **recessive**. See **monohybrid inheritance, backcross, incomplete dominance, codominance**.

dorsal Relating to features of, on, or near, that surface of an organism which is normally directed upwards, although in humans it is directed backwards. Compare **ventral**.

Down's syndrome A human abnormality caused by a **mutation** in which the **ovum** has an extra **chromosome**. Thus the resulting child has 47 chromosomes in each **cell** instead of the normal 46. This results in physical and mental retardation. See **amniocentesis**.

drug abuse The habitual consumption of substances which affect the **nervous system**. Such drugs include LSD, heroin and solvents (glue-sniffing). Overdosing on drugs can be fatal.

duodenum The first part of the mammalian **small intestine** leading from the **stomach** via the pyloric **sphincter**. The duodenum receives *pancreatic juice* from the **pancreas** and **bile** from the **liver**, and is an important digestive site.
The pancreatic juice contains **enzymes** which continue the **digestion** of food arriving from the stomach.
Bile contains *bile salts* which emulsify fat, forming small fat droplets, thus increasing the surface area available for **lipase** action.

Starch $\xrightarrow{\text{Amylase}}$ Maltose

Protein $\xrightarrow{\text{Trypsin}}$ Peptides $\rightarrow$ Amino acids

Fat $\xrightarrow{\text{Lipase}}$ Fatty acids + Glycerol

From the duodenum, the semidigested food is forced by **peristalsis** into the **ileum**.

ear The **organ** of hearing and balance in vertebrates. Hearing is a sensation produced by vibrations or sound waves which are converted into **nerve impulses** by the ear and transmitted to the **brain**.

Outer ear: the **pinna** is a funnel-shaped structure which directs sound waves into the ear and along the **auditory canal**, at the end of which is a very thin membrane,

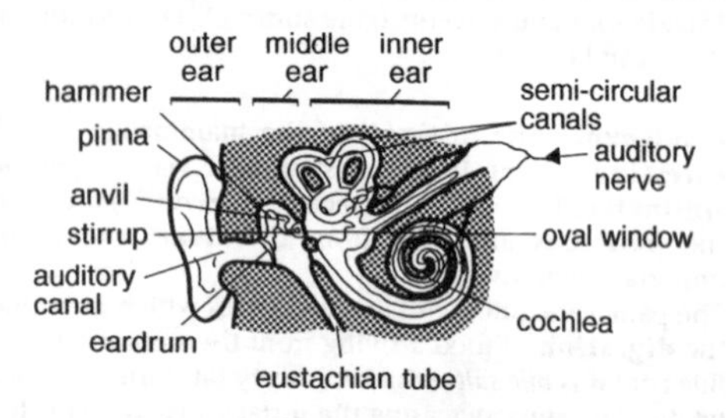

ear Structure of the human ear.

the *eardrum* (**tympanum**), which is made to vibrate by the sound waves.

Middle ear: this is an air-filled cavity connected to the back of the mouth (**pharynx**) by the **eustachian tube**, an arrangement which allows air into the middle ear ensuring equal air pressure on both sides of the eardrum.

Within the middle ear, there are three tiny bones, the **ossicles**, named by their shapes: *malleus* (hammer), *incus* (anvil), and *stapes* (stirrup).
The vibrations of the eardrum are transmitted through and amplified by the ossicles, the stapes finally vibrating against a membrane called the **oval window**, which separates the middle and inner ears.

Inner ear: this is fluid-filled and consists of the **cochlea** and **semicircular canals**.
The vibration of the stapes against the **oval window** sets up waves in the fluid of the cochlea. These waves stimulate **receptor cells** (*hair cells*) causing nerve impulses to be sent via the **auditory nerve** to the brain, where they are interpreted as sounds.

Balance is maintained by the semicircular canals in association with information received from the **eyes** and **muscles**. The semicircular canals contain fluid and receptor cells, which are stimulated by movements of the fluid during changes in posture. The nerve impulses initiated by these cells travel to the brain along the auditory nerve and trigger **responses** which cause the body to maintain normal posture.

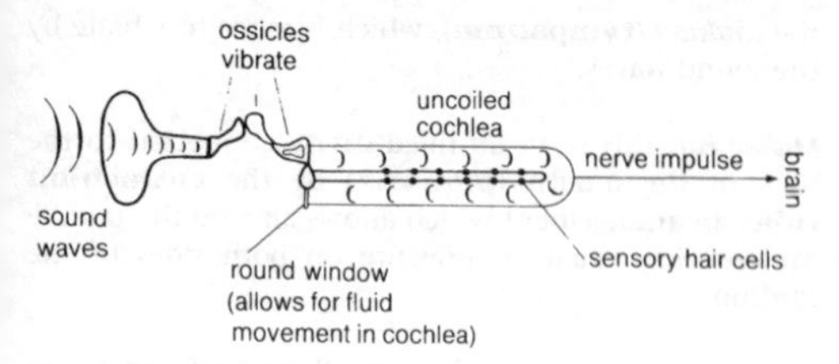

ear Sound waves are converted into nerve impulses.

ecdysis See **exoskeleton**.

ecology The study of the ways in which **communities** of plants and animals interact with one another and with their nonliving **environment**. See **ecosystem**.

ecosystem A **community** of organisms interacting with each other and with their nonliving **environment**, i.e. it is a natural unit consisting of living parts (plants and animals) and nonliving parts (light, water, air, etc.).

Habitat+Community→Ecosystem

Ecosystems can be lakes, ocean, forests, etc. The driving force behind all ecosystems is the flow of energy originating from the Sun.

ecosystem management The use of good management practices in order to conserve the **environment**, for example, by taking the appropriate measures to reduce **pollution**, controlling the use of **fertilizers**

and **pesticides** and adopting sensible methods of **agriculture** such as **crop rotation**.

ectoparasite A **parasite** living on the exterior of another organism, its *host*. Lice are examples of ectoparasites. See **endoparasite**.

ectotherm An animal whose body temperature varies with the temperature of its **environment**. All animals, excluding birds and mammals, are ectotherms, although they are often called *cold blooded*. Compare **endotherm**.

effector A specialized animal **tissue** or **organ** that performs a **response** to a **stimulus** from the **environment**. Examples are **muscles** and **endocrine glands**. See **sensitivity**.

electron See atom.

element A pure chemical that cannot be broken down into simpler substances. All the **atoms** of an element have the same number of **protons** or **electrons**. There are 92 naturally occurring elements.

embryo **1.** A young animal developed from a **zygote** as a result of repeated **cell division**. In mammals, the embryo develops within the female **uterus**, and in the later stages of **pregnancy** is called a **foetus**.
2. A young flowering plant developed from a fertilized **ovule**, which in *seed plants* is enclosed within a **seed**, prior to **germination**.

embryo sac The structure within the **ovules** of flowering plants in which the female **gametes** are located. See **carpel**.

endocrine glands or **ductless glands** Structures which release chemicals called **hormones** directly into the bloodstream in vertebrates and some invertebrates. The rate of secretion of hormones is often a response to changes in internal body conditions but may also be a response to changes in the **environment**. Compare **exocrine glands**. See diagram.

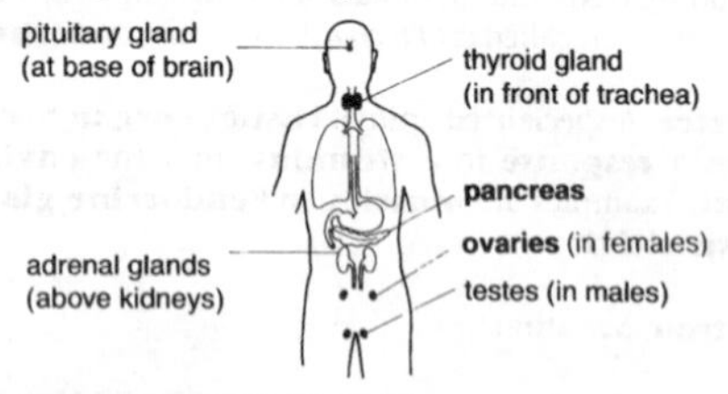

endocrine glands The main endocrine glands in the human body.

endoparasite A **parasite** living inside the body of another organism, its *host*. An example is the tapeworm. See **ectoparasite**.

endoskeleton or **internal skeleton** **The skeleton** lying within an animal's body, for example, the bony skeleton of vertebrates. Endoskeletons provide shape, support, and protection and work together with **muscles** to produce movement. Compare **exoskeleton**.

endotherm An animal which can maintain a constant narrow range of body temperature despite fluctuations

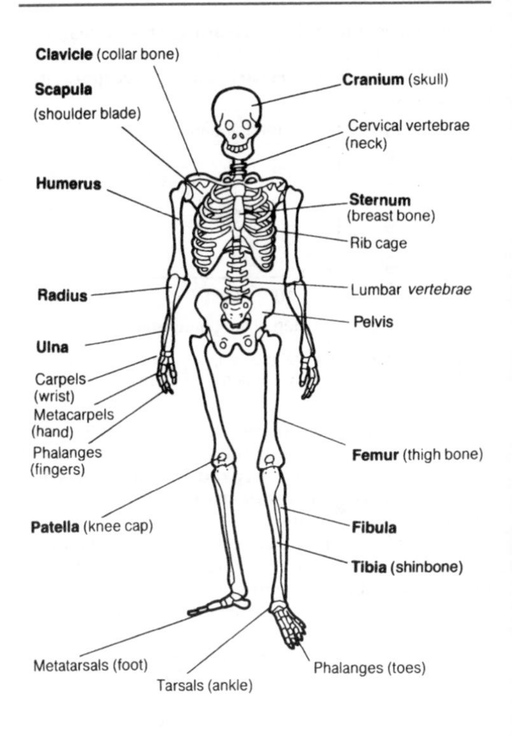

endoskeleton The human endoskeleton.

in the temperature of the **environment**. Mammals and birds are endotherms, although they are often called *warm blooded*. Compare **ectotherm**; see **temperature regulation**.

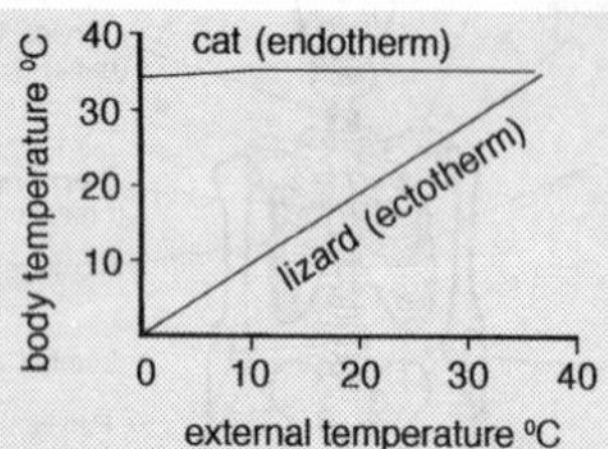

endotherm The relationship between external temperature and body temperature in an endotherm and an ectotherm.

energy The ability to do work. In living organisms that work is done in performing the seven characteristics of life: movement, feeding, **reproduction**, **excretion**, **growth**, **sensitivity**, **respiration**.
The types of energy are: heat, light, sound, electrical, chemical, nuclear, potential (stored), and kinetic (moving). Energy can be neither created nor destroyed, but it can be changed from one form into another. This scientific law is called the *conservation of energy*. Examples:

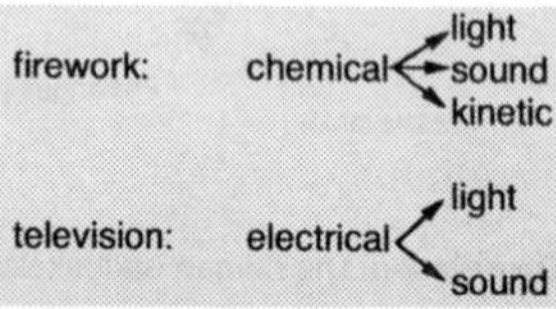

This concept of energy interconversion is important to living organisms, since green plants convert the light energy of sunlight into the chemical/potential energy of food, via the reaction of **photosynthesis**.

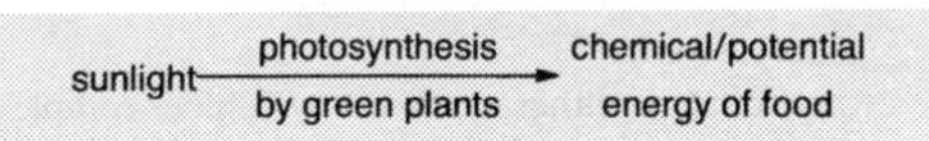

Other organisms can then release that chemical/potential energy via the reaction of **respiration** and convert it into other useful forms. For example:

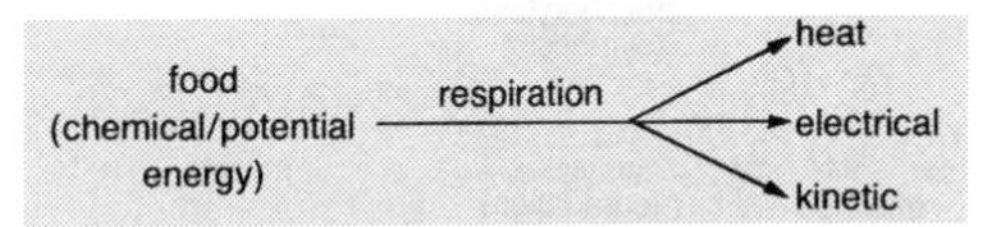

environment The surroundings in which organisms live and which influence the distribution and success of organisms. Many factors contribute to the environment, including (a) nonliving **abiotic factors**, e.g. temperature, light, **pH** etc., and (b) living **biotic factors**, e.g. **predators, competition**.
The interaction of these factors determines the conditions within **habitats**, and 'selects' the **communities** of organisms which are best suited to these conditions.

enzymes **Proteins** which act as *catalysts* within **cells**. Catalysts are substances which cause chemical reactions to proceed, and in cells there may be hundreds of reactions occurring, each one requiring a particular enzyme.

A + B —X→ no reaction

A + B —enzyme→ C + D

reactants products

Enzymes catalyse either *synthesis* by which complex compounds are formed from simple molecules, or *degradation* by which complex molecules are broken down to simple subunits by **hydrolysis**. See **digestion**.

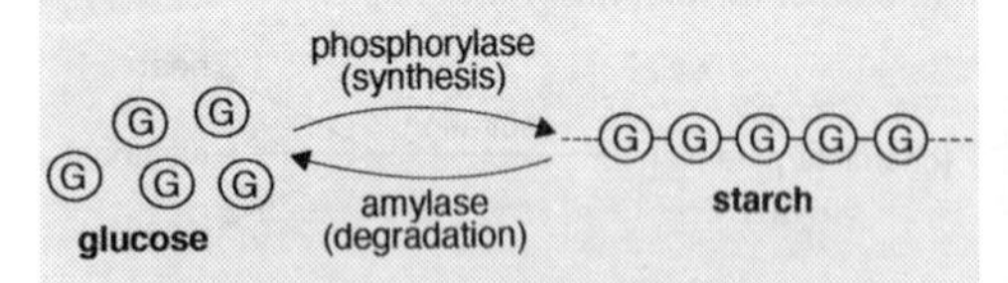

enzymes The synthesis and degradation of starch.

Enzyme characteristics

(a) Enzymes are proteins.

(b) Enzymes work most efficiently within a narrow temperature range. Thus human enzymes work best at 37 °C (body temperature) and this is called the

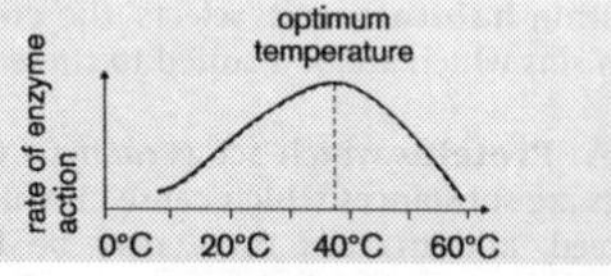

enzymes Graph showing the effect of temperature upon the rate of enzyme activity.

optimum temperature. Above and below this temperature their efficiency decreases, and at temperatures above 45 °C most enzymes are destroyed (**denaturation**).

(c) Enzymes have a **pH** at which they work most efficiently; this is known as the *optimum pH*. For example, the **saliva** enzyme salivary amylase works best at neutral or slightly acid pH. The **stomach** enzyme **pepsin** will only function in an acid pH, while the enzyme **trypsin** in the **intestine** favours an alkaline pH.

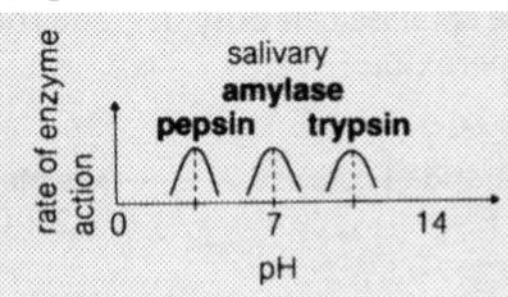

enzymes Graph showing the effect of pH upon enzyme activity.

(d) The rate of an enzyme-catalysed reaction increases as the enzyme concentration increases.

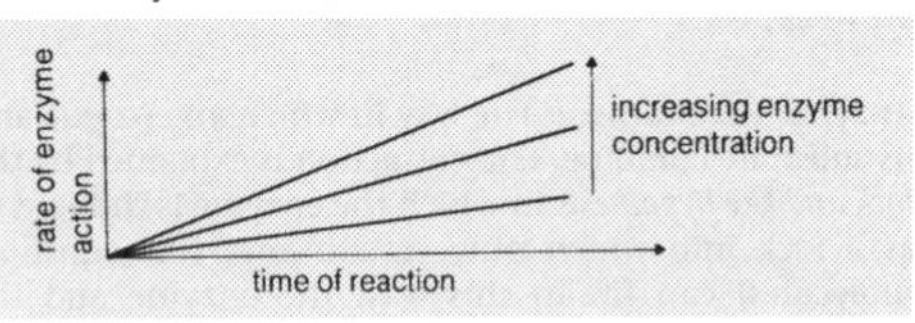

enzymes Graph showing the effect of enzyme concentration upon the rate of reaction.

(e) The rate of an enzyme-catalysed reaction increases as the **substrate** concentration increases, up to a maximum point.

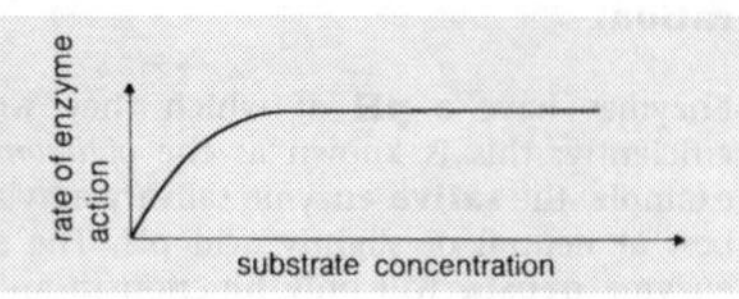

(f) Normally an enzyme will catalyse only one particular reaction, a property called *specificity*. For example, the enzyme catalase can only degrade the compound hydrogen peroxide.

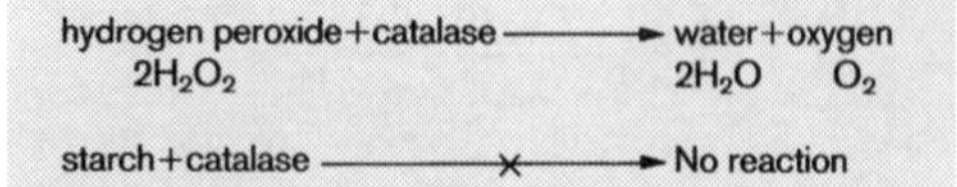

Most enzymes are named by adding the suffix *-ase* to the name of the enzyme's substrate (or by changing *-ose* to *-ase*). For example, **maltase** acts on maltose; urease acts on **urea**, etc.

enzyme mechanism The way in which an enzyme and its substrate combine. Enzyme action is explained by the *lock and key hypothesis* in which the enzyme is thought of as a lock into which only certain keys (the substrate molecules) can fit. In this way, the enzyme and the substrate are brought together and the reaction can occur.

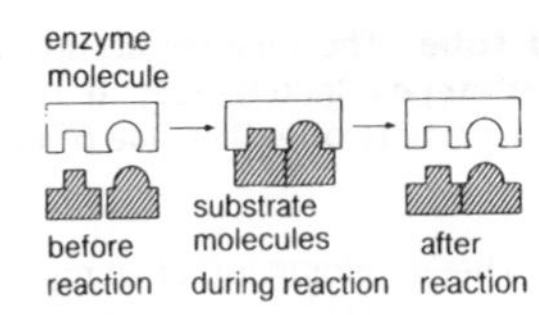

enzyme mechanism Sequence illustrating an enzyme-catalysed synthesis. Reversing the order of the diagrams shows how an enzyme-catalysed degradation occurs.

epidermis The protective outermost layer of **cells** in an animal or plant. The epidermis of many **multicellular** invertebrates is one cell thick and is often covered with a **cuticle**. In most vertebrates the epidermis is the outer layer of **skin**, and in land vertebrates may have several layers of dead cells. In plants, the epidermis is one cell thick, and on aerial (above-ground) structures may have a cuticle. See **leaf**, **root**, **stem**.

epiglottis The flap of **cartilage** and membrane at the base of the tongue on the ventral wall of the **pharynx**. It closes the **trachea** during swallowing. See **digestion**.

epithelium Lining **tissue** in vertebrates consisting of closely packed layers of **cells**, covering internal and external surfaces. Examples are the **skin** and the lining of the **breathing**, digestive and urinogenital **organs**. Epithelia may also contain specialized structures, e.g. **cilia**, **goblet cells**.

erythrocyte See **red blood cell**.

eustachian tube The tube connecting the **middle ear** to the **pharynx** in tetrapods. It is important in equalizing air pressure on either side of the **eardrum**. See **ear**.

evolution The development of complex organisms from simpler ancestors, occurring over successive generations. See **natural selection**.

excretion Elimination of the waste products of **metabolism** by living organisms. The main excretory products are water, carbon dioxide and nitrogenous compounds, e.g. **urea**. In simple organisms excretion occurs through the **cell membrane** or **epidermis**, in higher plants via the **leaves**, while most animals have specialized excretory **organs**. For example, in man the **lungs** excrete water and carbon dioxide, and the **kidneys** excrete urea.

exercise Any physical activity or bodily exertion which contributes to good health. Evidence suggests that regular exercise reduces the risk of **heart** disease, probably as a result of improved **blood** circulation. See **circulatory system**.

exocrine glands Structures in vertebrates which release secretions to **epithelial** surfaces via **ducts**. Examples are sweat glands; salivary glands. Compare **endocrine glands**.

exoskeleton or **external skeleton** A **skeleton** lying outside the body of some invertebrates, for example, the **cuticle** of insects and the shells of crabs. Some organisms shed and renew their exoskeletons periodically to allow **growth**, a process known as *moulting* or **ecdysis**. Compare **endoskeleton**.

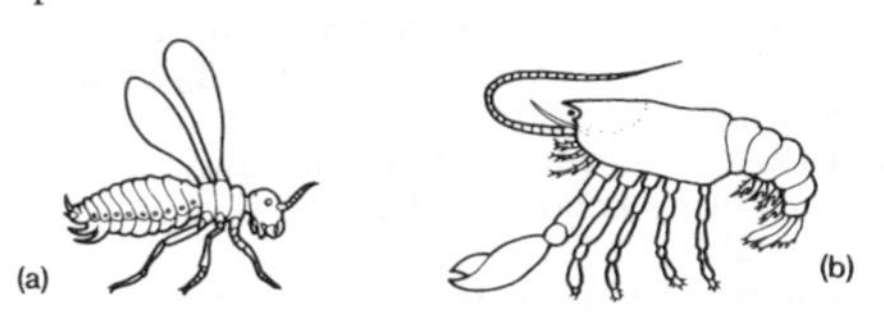

exoskeleton The exoskeletons of (a) an insect and (b) a crustacean.

eye A **sense organ** that responds to light, ranging from very simple structures in invertebrates to the complex organs of insects and vertebrates. Eye muscles enable the eye to move up and down and from side to side.
The **sclerotic** is a tough protective layer which at the front of the eye forms the transparent **cornea**.

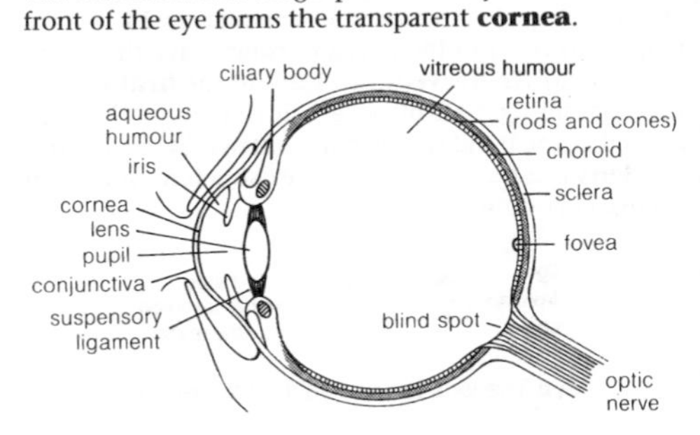

eye Vertical section through the human eye.

The **choroid** is a black-pigmented layer under the sclerotic, rich in **blood vessels** supplying food and oxygen to the eye.
The **retina** is a layer of **neurones** which are sensitive to light. There are two types of cell in the retina, named by their shape:
(a) **rods** are very sensitive to low intensity light and are particularly concentrated in the eyes of nocturnal animals;
(b) **cones** are sensitive to bright light. There are different types which are stimulated by different wavelengths of light and are thus responsible for *colour vision*. Animals whose retinas lack cones are colour blind, while human colour blindness is caused by a defect in the cones.

The **fovea** (*yellow spot*) is a small area of the retina containing only cones in great concentration and giving the greatest degree of detail and colour.
The **blind spot** is that part of the retina at which nerve fibres connected to the rods and cones leave the eye to enter the **optic nerve** which leads to the **brain**. Since there are no light-sensitive cells at this point, an image formed at the blind spot is not registered by the brain.
The **lens** is a transparent biconvex structure which can change curvature.

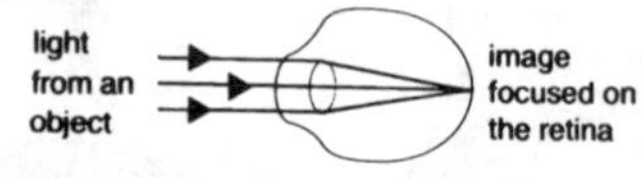

eye The lens focuses light on the retina.

The lens is held in place by **suspensory ligaments**

which are attached to the **ciliary muscles**, the contraction or relaxation of which alters the shape of the lens, allowing both near and distant objects to be focused sharply. This is called **accommodation**.

The **iris** is the coloured part of the eye, containing **muscles** which vary the size of the *pupil*, the hole through which light enters the eye. In poor light, the pupils are wide open (dilated), to increase the brightness of the image. In bright light the pupils are contracted to protect the retina from possible damage. This mechanism is an example of a **reflex action**.

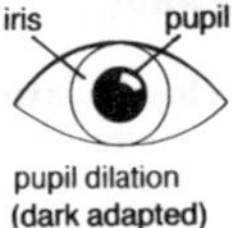

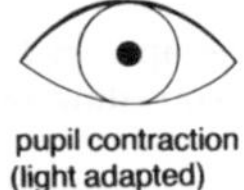

eye The pupil dilates and contracts according to the intensity of the light.

Because the pupil is small, light rays enter the eye in such a way that the image at the retina is upside down (inverted). This inversion of the image is corrected by the brain.

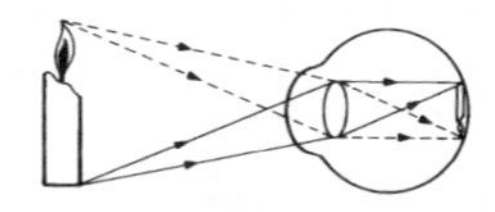

eye The image is smaller than the object, and is inverted.

The **aqueous humour** and **vitreous humour** are

fluids which fill the chambers of the eye. They help to maintain shape, focus light, and allow nutrients, oxygen, and wastes to diffuse to and from the eye cells.

F_1 generation (first filial generation) The first generation of **progeny** obtained in breeding experiments. Successive generations are called F_2 etc. See **monohybrid inheritance**.

faeces The solid or semisolid remains of undigested food, **bacteria**, etc, which are formed in the **colon** of vertebrates and expelled via the **anus**.

family The unit used in the **classification** of living organisms consisting of one or more genera (singular **genus**).

fats or **lipids** **Organic compounds** containing the elements carbon, hydrogen, oxygen. Fats consist of three *fatty acid* molecules (which may be the same or different) bonded to one *glycerol* molecule.
Fat deposits under the **skin** act as long term **energy** stores, yielding 39 kJ/g when respired; these deposits also provide heat insulation.
Fat is an important constituent of the **cell** membrane

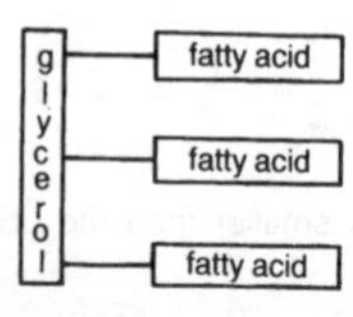

fats Structure of a fat molecule.

(see **selectively permeable membrane**) and its insolubility in water is utilized in the waterproofing systems of many organisms. See **respiration**.

feedback regulation See **insulin**.

femur **1.** The part of an insect limb nearest to the body. **2.** The thighbone of **tetrapod** vertebrates. See **endoskeleton**.

fermentation The degradation of **organic compounds** in the absence of oxygen for the purpose of **energy** production, by certain organisms, particularly **bacteria** and **yeasts**. Fermentation is a form of anaerobic **respiration**. See **brewing**.

$$\underset{C_6H_{12}O_6}{\text{glucose}} \xrightarrow{\text{yeast}} \underset{2C_2H_5OH}{\text{ethanol}} + \underset{2CO_2}{\text{carbon dioxide}}$$

$$\text{ADP} \longrightarrow \textbf{ATP}$$

fermentation The fermentation of sugars by yeast produces the by-products ethanol and carbon dioxide.

fertilization The fusing of **haploid gametes** during **sexual reproduction** resulting in a single cell, the **zygote**, containing the **diploid** number of chromosomes. It occurs in both animals and plants. In animals, fertilization is either external or internal.

External fertilization occurs when the gametes are passed out of the parents and fertilization and development take place independently of the parents. External fertilization is common in aquatic organisms where the movement of water helps the gametes to meet.

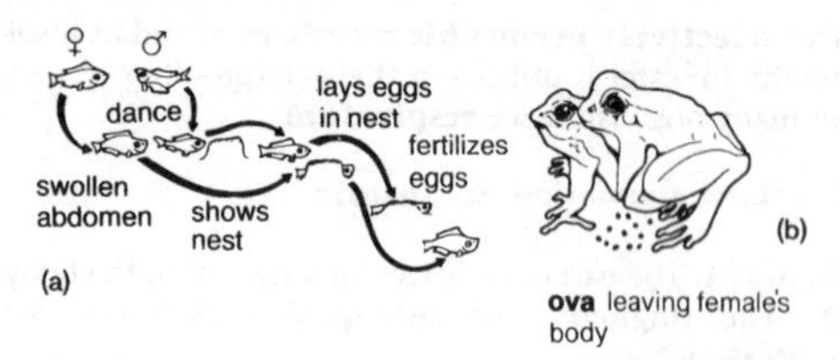

fertilization External fertilization in (a) fish: the stickleback; and (b) amphibians: the toad.

Internal fertilization is particularly associated with terrestrial animals, for example, insects, birds and mammals, and involves the union of the gametes within the female's body. The advantages of internal fertilization are (a) the **sperms** are not exposed to unfavourable dry conditions, (b) the chances of fertilization occurring are increased, and (c) the fertilized **ovum** is protected within a shell (birds) or within the female body (mammals).

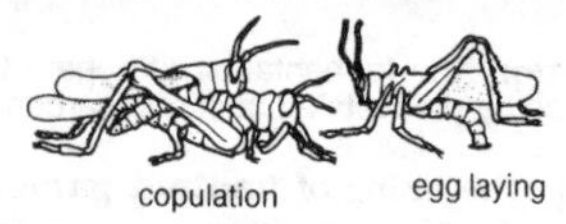

fertilization Internal fertilization in an insect: the locust.

Sperm cells, produced in the **testes**, are passed out of the **penis** during **copulation** in which the penis is inserted in the **vagina**. The sperms move through the **uterus** and, if an ovum is present in an **oviduct**, fertilization can occur there.

Fertilization of ovum. The fertilized ovum (zygote) con-

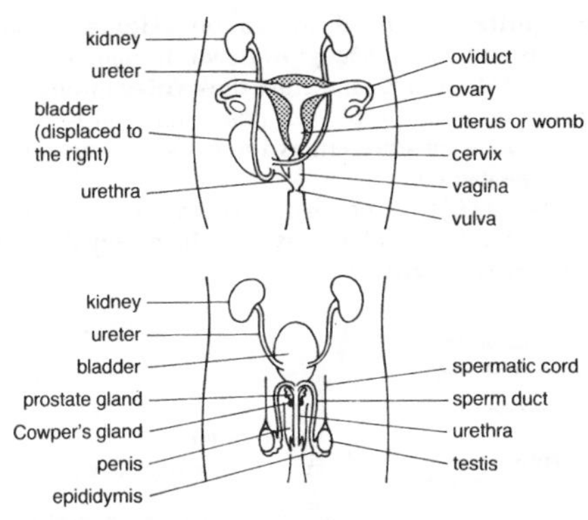

fertilization The human reproductive organs.

tinues moving towards the uterus, dividing repeatedly as it does so. On arrival at the uterus, the zygote, by now a ball of cells, becomes embedded in the prepared wall of the uterus. This is called **implantation** and further development of the **embryo** occurs in the uterus. See **pregnancy**, **birth**.

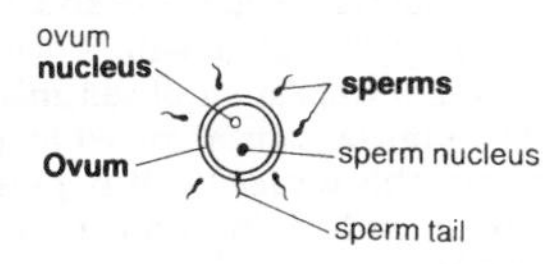

fertilization Fertilization of the ovum.

Fertilization in plants. In flowering plants, after **pollina-**

tion, **pollen** grains deposited on **stigmas** absorb nutrients and *pollen tubes* grow down through a narrow region called the *style* and enter the **ovules** through the **micropyles**. The tip of each pollen tube breaks down and the male **gamete** enters the ovule and fuses with the female gamete.

After fertilization, the ovule, containing the plant embryo, develops into a **seed**, and the **ovary** develops into a **fruit**. See **flower**.

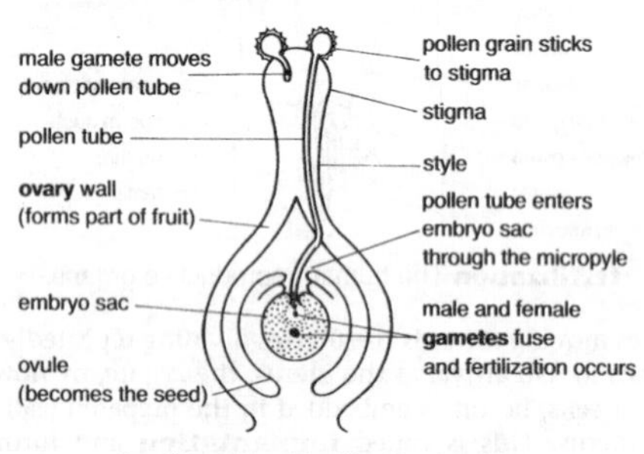

fertilization Fertilization in plants.

fertilizer Any substance added to **soil** to increase the quantity or quality of plant growth. When crops are harvested, the natural circulation of soil **mineral salts** is disturbed, i.e., mineral salts absorbed by plants are not returned to soil. This is called **soil depletion** and may render the soil infertile. Fertilizers replenish the soil and are of two types:

(a) organic fertilizers such as compost, sewage;

(b) inorganic fertilizers such as ammonium sulphate.

See **organic compounds**, **inorganic compounds**.

fibrinogen A soluble **plasma protein** involved in **blood clotting**.

fibula The posterior of two **bones** in the lower hind-limb of **tetrapods**. In humans it is the outer bone of the leg below the knee. See **endoskeleton**.

fitness The state of being in healthy physical condition. Fitness is achieved and maintained by adopting a healthy lifestyle, for example by having a **balanced diet**,taking regular **exercise**, and avoiding alcohol, smoking and drug abuse.

flagellum A microscopic motile thread projecting from certain **cell** surfaces and causing movement by lashing back and forth. Flagella are usually larger than **cilia**, and less numerous, and are responsible for locomotion in many **unicellular** organisms and reproductive cells.

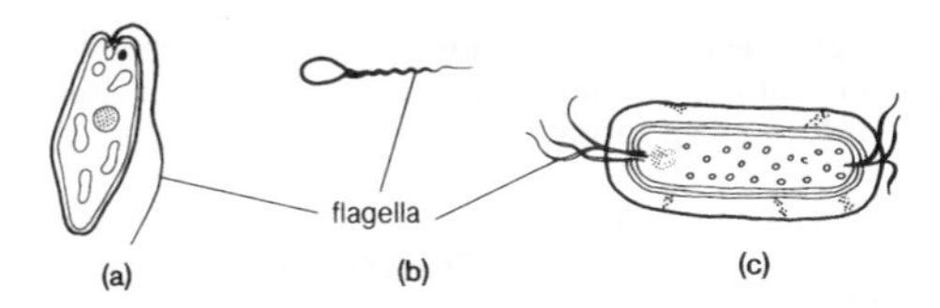

flagellum Flagella in (a) *Euglena*, (b) motile sperm and (c) motile bacterium.

flower The organ of **sexual reproduction** in flowering plants (angiosperms).

Insect-pollinated flowers have brightly coloured and scented petals, and usually have a *nectary* at the base of the flower. The nectary contains a sugar solution called *nectar*. The **stamens** and **carpels** (with sticky **stigmas**) are within the flower. These adaptations favour insect pollination.
Wind-pollinated flowers are small, often green and unscented, and do not have nectaries. The **anthers** and feathery stigmas dangle out of the flowers when ripe thus facilitating wind pollination. See **fertilization, pollen, pollination**.

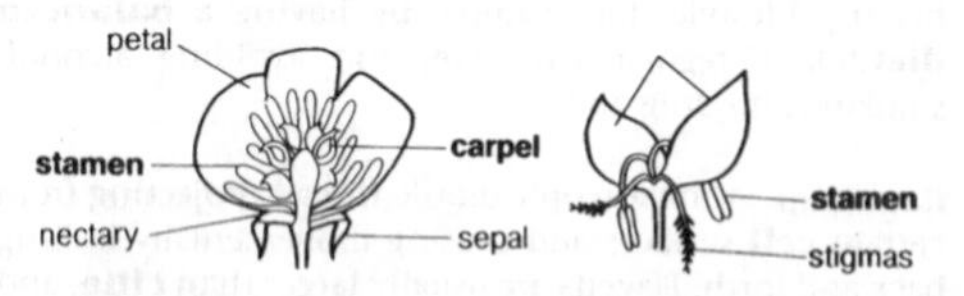

flower The structure of (a) and insect-pollinated flower and (b) a wind pollinated flower.

foetus The mammalian **embryo** after development of main features. In humans this is after about three months of **pregnancy**.

follicle-stimulating hormone (FSH) A **hormone** secreted by the vertebrate **pituitary gland**. See **ovulation**.

food calorimeter A device for measuring the **energy** content of food.
A weighed food sample is ignited using the heating filament. In the presence of oxygen, the food sample is

food	energy content
protein	17 kJ/g
carbohydrate	17 kJ/g
fat	39 kJ/g

food calorimeter The energy content of the different kinds of food.

completely combusted, the released energy being transferred to the surrounding water, which shows a rise in temperature.

From this rise in temperature, it is possible to calculate the energy content of the food in **kilojoules** per gram (kJ/g).

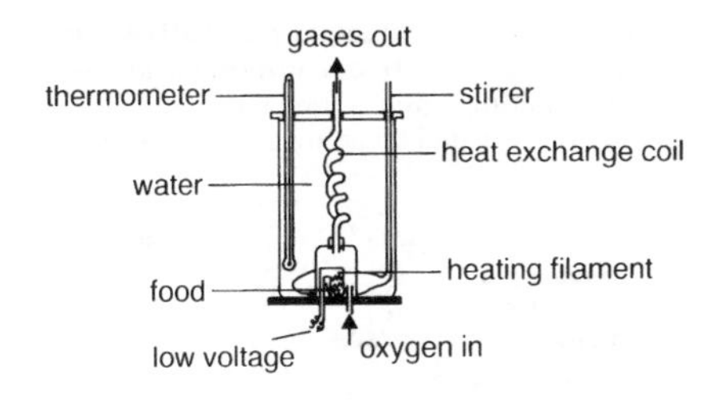

food calorimeter

Note: the kilojoule has replaced the calorie as the unit of energy.

food capture The method by which organisms obtain food. Many **heterotrophic** organisms have developed very specialized methods and structures for obtaining food; a variety of examples is given below.

(a) *Mammals without teeth.* Anteaters have a long and sticky tongue for catching ants. Blue whales have modified mouth parts to filter **plankton** out of water.

(b) *Filter feeding.* Like the blue whale, many aquatic organisms, especially invertebrates, filter plankton.

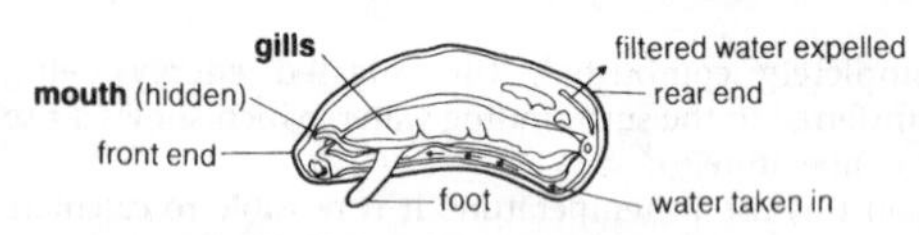

food capture Filter feeding: *Mytilus* (the edible mussel) shown with one shell removed.

(c) *Feeding by sucking.* Houseflies pass **saliva** out onto their food, e.g. sugar. **Digestion** begins immediately and the resulting liquid is then taken in by a sucking pad on an elongated structure called a *proboscis*.

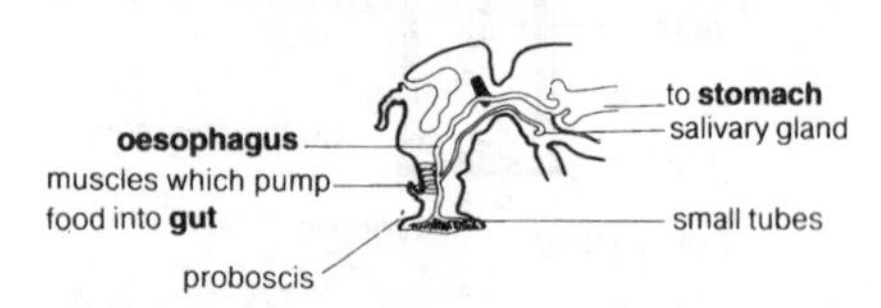

food capture The housefly sucks up dissolved food.

Female mosquitoes pierce the human skin, inject a fluid which prevents **blood clotting**, and then suck up some blood. This feeding mechanism can cause the disease malaria, since the **parasite** involved may be transmitted during feeding.

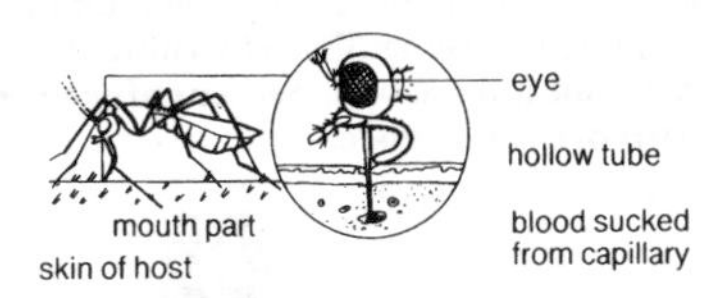

food capture The female mosquito sucks blood.

Butterflies feed on the nectar produced by flowering plants. They suck the nectar from the **flower** by means of a long tube-like proboscis, which remains coiled when not in use.

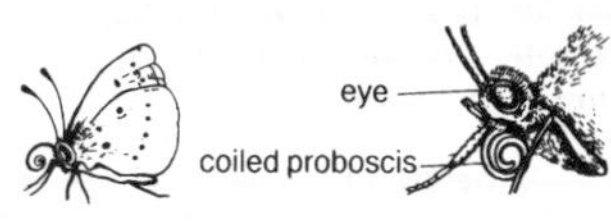

food capture The buttefly's long, coiled proboscis.

Greenflies use a long piercing proboscis to suck plant

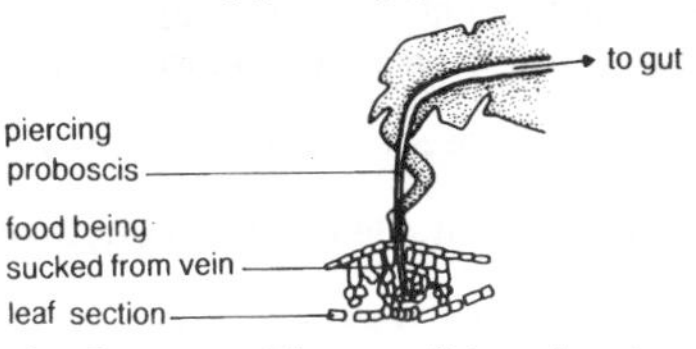

food capture The greenfly's proboscis.

juices from **leaves** and **stems**.
(d) *Biting without teeth.* Locusts eat their own weight of plant material every day. They have powerful biting jaws called *mandibles* which have very hard biting edges, which are brought together during feeding in a precise and efficient shearing action. See **carnivore**, **dentition**, **herbivore**, **omnivore**.

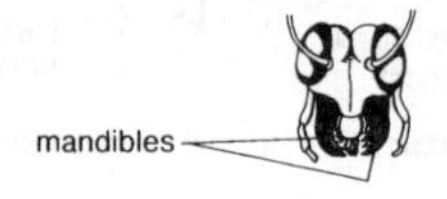

food capture The locust's biting jaws.

food chain A food relationship in which **energy** and carbon compounds obtained by green plants via **photosynthesis** are passed to other living organisms, i.e., plants are eaten by animals which in turn are eaten by other animals and so on.

green plant (producer) → **herbivore** (primary consumer) → small **carnivore** (secondary consumer) → large carnivore (tertiary consumer)

The arrows indicate 'is eaten by' and the direction of energy flow. An example of such a food chain is:

plants → insects → lizards → snakes

Not all food chains are as long as the above, for example:

grass → sheep → man
grass → antelope → lion

Such simple food chains seldom exist independently; more often several food chains are linked in a more complicated relationship called a *food web*.

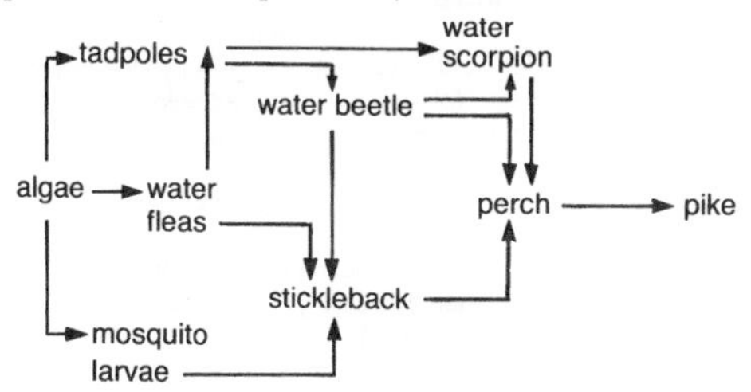

food chain Part of the food web in a freshwater pond.

All food webs are delicately balanced. Should one link in the web be destroyed, all the other organisms will be affected. For example, in the pond food web, if the perch disappeared as the result of disease, the pike population would decrease while the water scorpions would increase.

food consumers **Heterotrophic** organisms which, after green plants, occupy the subsequent links in a **food chain**.

food producers **Autotrophic** organisms, mainly green plants, which occupy the first level in a **food chain**.

food tests Chemical tests used to identify the components of a food sample. Some common food tests are shown below.

protein	+	Biuret reagent (blue)	→	violet/purple colour
reducing sugar	+	Benedict's reagent (blue)	heat →	green/yellow/brick red colour
starch	+	iodine solution (brown)	→	blue/black colour
fat + water	+	ethanol (clear)	→	white emulsion
vitamin C	+	dichlorophenolindo-phenol (DCPIP) (blue)	→	DCPIP (clear)

food web See **food chain**.

fossil The mineralized remains of animal and plant **tissue** from previous geological ages, which have been preserved in the Earth's crust. The study of fossils is called **palaeontology**, and has supplied important evidence for **evolution**.

fossil fuels The fossilized remains of plants and animals which, over millions of years, have been transformed into oil, coal and natural gas. See **carbon cycle**, **greenhouse effect**, **acid rain**.

fovea or **yellow spot** The area of the **retina** in some vertebrate **eyes**, specialized for acute vision. It contains numerous **cones** but no **rods**.

fruit The ripened **ovary** of a **flower**, enclosing **seeds**, formed as the result of **pollination** and **fertilization**.

The fruit protects the seed and aids its dispersal.

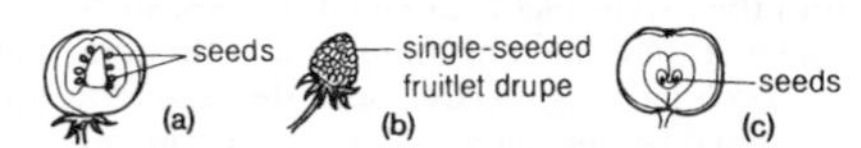

fruit (a) tomato (b) blackberry (c) apple.

fruit and seed dispersal The methods by which most flowering plants spread seeds far away from the parent plant, thus (a) avoiding **competition** for resources and (b) ensuring wide colonization, so that suitable **habitats** are likely to be encountered by a proportion of seeds. The methods are:

(a) *Wind dispersal*: Air currents carry the fruits or seeds which usually show an adaptation to increase their surface area.

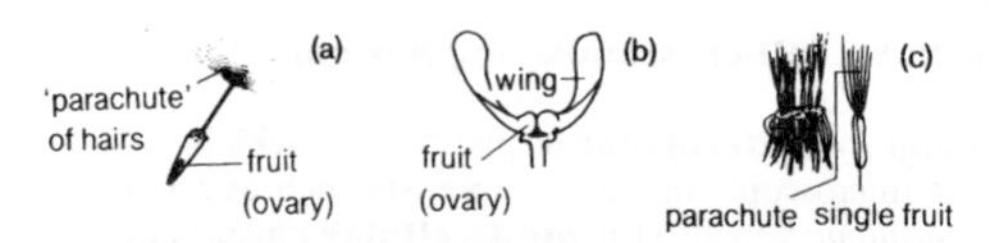

fruit and seed dispersal Methods of wind dispersal: (a) dandelion (b) sycamore (c) groundsel.

(b) *Animal dispersal*: Hooked fruits, e.g. burdock, stick to

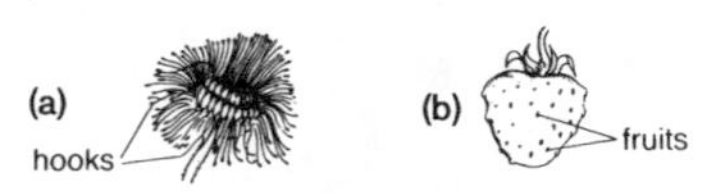

fruit and seed dispersal Examples of fruit dispersed by animals (a) burdock (b) strawberry.

animals' coats and may be brushed off some distance from the parent plant. Succulent berries, such as the strawberry, are eaten by animals, and the small hard fruits containing seeds pass through the **gut** unharmed before being released in the **faeces**.

(c) *Explosive dispersal*: Unequal drying of part of a fruit causes the fruit to burst, shooting the seeds away from the parent plant.

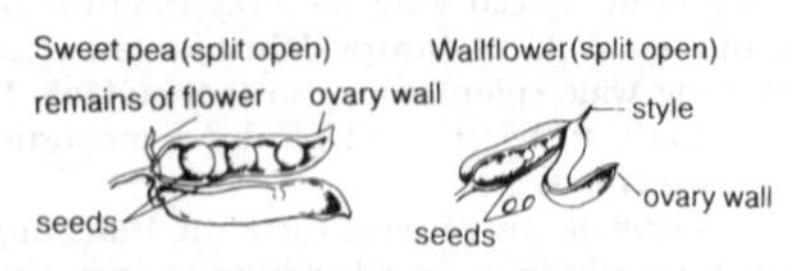

fruit and seed dispersal Sweet pea and wall-flower disperse their seeds by explosion of the fruit.

FSH See **follicle-stimulating hormone**.

fungi A **heterotrophic** plant group which includes the microscopic moulds and **yeasts** such as *Mucor* and *Penicillium* and also the **multicellular** mushrooms and toadstools. Some fungi are **parasites** causing plant disease such as potato blight and Dutch elm disease. In humans, fungi cause athletes foot, ringworm and **lung**

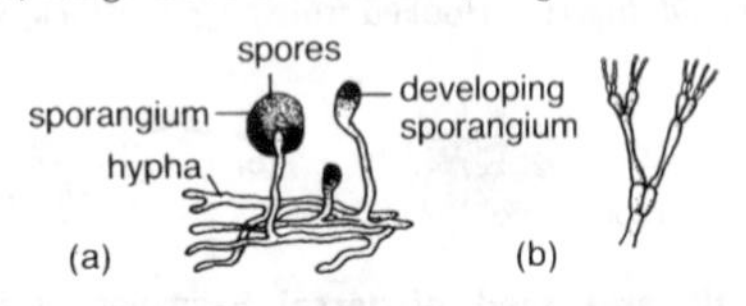

fungi (a) Mucor (b) Penecillium.

infections. Other fungi are useful to man, for example as a source of **antibiotics** and in **brewing**.

gall bladder A small bladder in or near the vertebrate **liver**, in which **bile** is stored. When food enters the **intestine**, the gall bladder empties bile into the **duodenum** via the **bile duct**. See **digestion**.

gamete A reproductive **cell** whose nucleus is formed by **meiosis** and contains half the normal **chromosome** number (i.e. they are **haploid** cells). Human male gametes are **spermatozoa** and human female gametes are **ova** (egg cells). Male and female gametes fuse during **fertilization** forming a **zygote** in which the normal chromosome number (**diploid**) is restored.

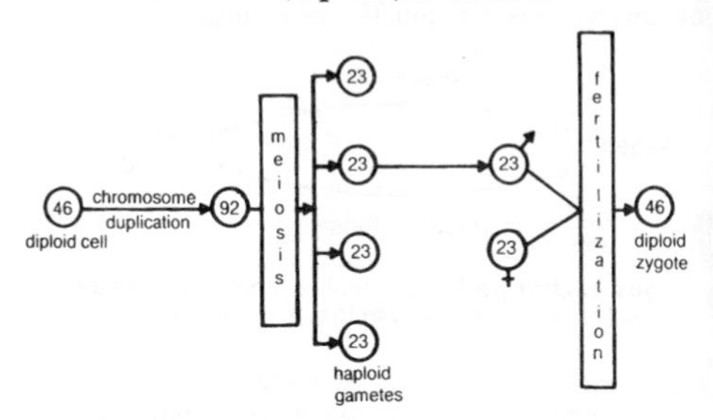

gamete In humans, the haploid gametes have 23 chromosomes and the diploid zygote has 46.

gas exchange The process by which organisms exchange gases with the **environment** for the purpose of **metabolism**. Most organisms require a continuous supply of the gas oxygen for the reaction of **respiration**:

> **glucose** + oxygen → **energy** + carbon dioxide + water

In addition, green plants require carbon dioxide for the reaction of **photosynthesis**:

> carbon dioxide + water —(light energy, chlorophyll)→ **carbohydrate** + oxygen

Both reactions use and produce gases which are interchanged between the atmosphere (in the case of land organisms) or water (aquatic organisms).

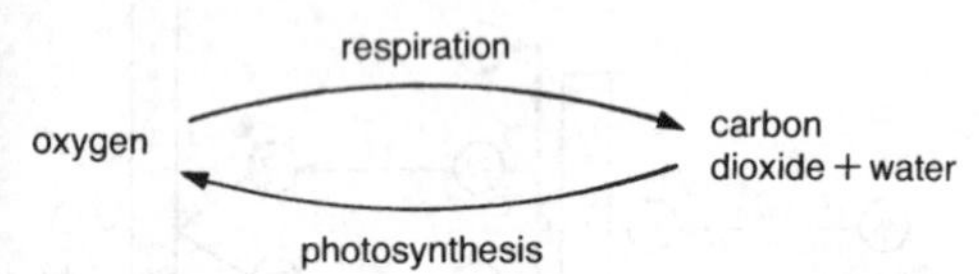

gas exchange Both repiration and photosynthesis involve gas exhange with the environment.

Gas-exchange surfaces. Gas exchange takes place across surfaces which have the following characteristics:
(a) a large surface area for maximum gas exchange;
(b) the surface is thin to allow easy **diffusion**;
(c) the surface is moist since gas exchange occurs in solution;
(d) in animals, the surface has a good blood supply, since

the gases involved are transported via the **blood**. Gas exchange in fish occurs across **gills** which consist of *gill arches* to which are attached numerous gill filaments. Water is taken in via the mouth and passed over the gills where oxygen dissolved in the water is absorbed into **blood capillaries** while carbon dioxide diffuses into the water.

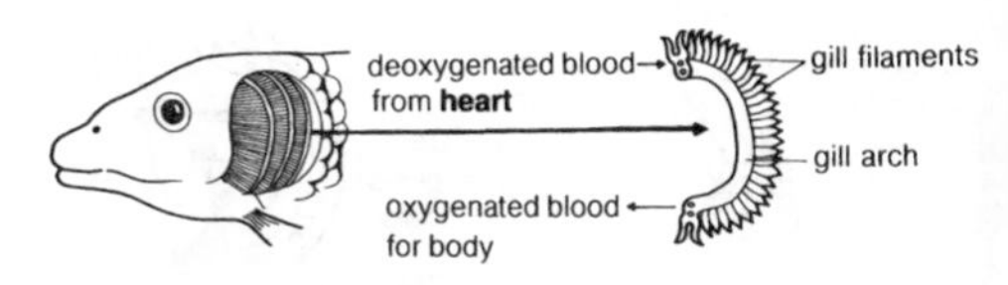

gas exchange The position and structure of the gills of fish.

Gas exchange in insects: air enters insects through pores called **spiracles** and is carried through a branching system of **tracheae** and thus into smaller branches called *tracheoles* which are in contact with the **tissues**.

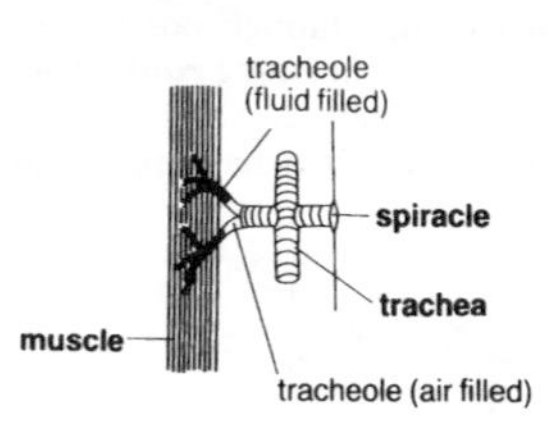

gas exchange Gas exchange in insects occurs via the fluid in the tracheoles.

Gas exchange in mammals occurs as the result of *concentration gradients* (see **diffusion**) existing between the air in the alveoli and the deoxygenated **blood** arriving from the **heart**. These gradients cause diffusion of oxygen from the alveoli into **red blood cells** and diffusion of carbon dioxide from the blood into the alveoli. See **breathing, lungs**.

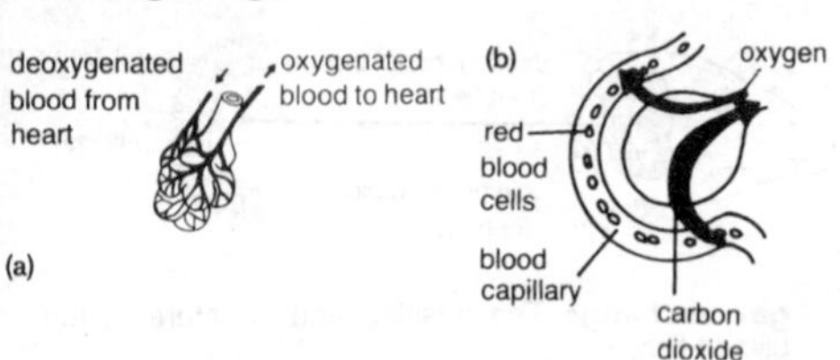

gas exchange (a) Alveoli and associated blood vessels. (b) Gas exchange in the alveolus.

Gas exchange in plants:

(a) *Terrestrial plants*: **gas exchange** in **leaves** and young **stems** occurs through pores in the **epidermis** called **stomata**. In young **roots** it occurs by diffusion between the roots and air in the **soil**. In older stems and roots where bark has formed, gas exchange

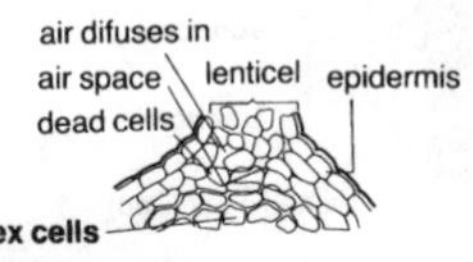

gas exchange Section throught plant stem showing lenticel.

takes place through gaps in the bark called **lenticels**.
(b) *Aquatic plants*: submerged plants, e.g. pond-weed, have no stomata, gas exchange occurring by diffusion across the **cell membranes**. Aquatic plants with floating leaves, e.g. water lily, have stomata only on the upper leaf surface.
(c) *Nongreen plants*: mushrooms, for example, carry out **respiration** but not **photosynthesis**. Gas exchange occurs by diffusion between the plant cells and the surrounding air.

gastric Relating to parts and functions of the body connected with the **stomach**.

genes The subunits of **chromosomes** consisting of lengths of **DNA** which control the hereditary characteristics of organisms. Genes consist of up to one thousand *base pairs* in a DNA molecule, the particular sequence of which represents coded information. This is known as the **genetic code** and determines the types of **proteins** synthesized by **cells**, particularly **enzymes**, which then dictate the structure and function of cells and **tissues**, and ultimately organisms, i.e., a cell or an organism is an expression of the genes it has inherited (and the **environment** in which it lives).
The genetic code is the arrangement of nitrogen base pairs in DNA. Each group of three adjacent base pairs (*triplets*) is responsible for linking together, within the cell, **amino acids** to form protein. The sequence, types and numbers of amino acids determine the nature of the proteins, which in turn determine the characteristics of cells. For example, the base triplet GTA codes for the amino acid histidine while GTT codes for glutamine.

Consider two fruit flies (*Drosophila*), one with a gene X controlling body colour, while the other's body colour is controlled by gene Y.

gene X —synthesizes→ enzyme X —catalyses→ pigment X
gene Y —synthesizes→ enzyme Y —catalyses→ pigment Y
pigment X —produces→ light body
pigment Y —produces→ dark body

genetic code See **genes**.

genetic engineering The transfer of pieces of **chromosome** from one organism to another. For example,

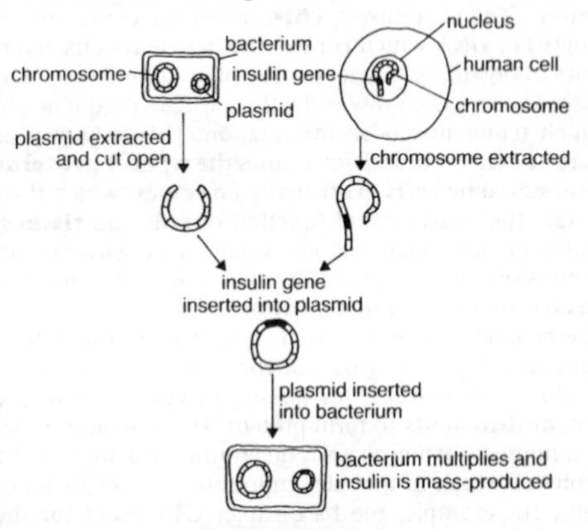

genetic engineering

the **gene** for human **insulin** can be inserted into a **bacterium** which will then produce insulin that can be isolated and purified for treatment of diabetes. See diagram.

genetics The study of *heredity*, which is the transmission of characteristics from parents to offspring via the **genes** in the **chromosomes**. Heredity is investigated by performing breeding experiments and then comparing the characteristics of the parents and offspring. Such experiments were first done by Gregor Mendel in the 1860s using pea plants. See **monohybrid inheritance**, **backcross**, **incomplete dominance**, **codominance**.

genotype The genetic composition of an organism, i.e. the particular set of **alleles** in each cell. In breeding experiments, genotypes are represented by symbols, capital letters denoting the **dominant** alleles and small letters denoting the **recessive** alleles. See **monohybrid inheritance**.

genus A unit used in the **classification** of living organisms, consisting of a number of similar **species**.

geotropism A form of **tropism** relative to gravity. Plant **shoots** grow away from gravity (*negative geotro-*

geotropism

pism), but most **roots** grow downwards and thus show *positive geotropism.*

germination The beginning of **growth** in **spores** and **seeds**, which normally proceeds only under certain environmental conditions, for example, the availability of water and oxygen, and a favourable temperature. If these conditions are not present, spores and seeds may remain alive for some time without germinating. In this state, known as *dormancy*, the seeds are *dormant*, i.e. inactive while they await the right conditions.
Seed germination in flowering plants: there are two types of germination: *hypogeal* and *epigeal*. They are distinguished by what happens to the **cotyledons** during development of the seedling. In hypogeal germination, the

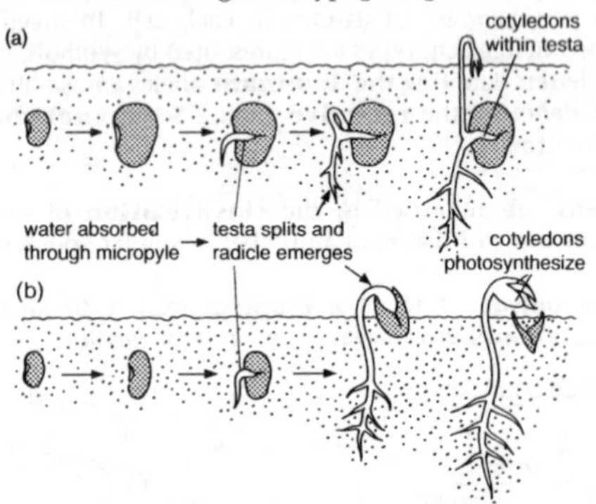

germination (a) Hypogeal germination (broad bean). (b) Epigeal germination (French bean).

cotyledons remain below ground while in epigeal germination the cotyledons are taken above ground. In both cases water is absorbed via the **micropyle**, the **testa** splits open and the **radicle** emerges. See diagram.

gestation period See **pregnancy**.

gills The **gas-exchange** surface of aquatic animals. In fish, gills are usually internal, projecting from the **pharynx**, while in amphibian **larvae**, they are external.

gland A cell or **organ** that synthesizes chemical substances and secretes them into the body either through a duct or direct to the bloodstream. See **endocrine gland**.

glomerulus (*pl.* **glomeruli**) The knot of **blood capillaries** within the **Bowman's capsule** of the mammalian **kidney**.

glottis The opening of the **larynx** into the **pharynx** of vertebrates.

glucose A **monosaccharide carbohydrate**, synthesized in green plants during **photosynthesis**, and serving as an important **energy** source in animal and plant **cells**. See **monosaccharide**, **respiration**.

glycogen A **polysaccharide carbohydrate** consisting of branched chains of **glucose** units, which is important as an **energy** store in animals. In vertebrates, glycogen is stored in **muscle** and **liver cells** and is readily converted to glucose by **enzymes**. See **insulin**.

goblet cells Specialized **cells** in certain **epithelia**, which synthesize and secrete **mucus**. Goblet cells are common in vertebrates, and are found in, for example, the intestinal and respiratory tracts of mammals. See **intestine**.

goblet cells Goblet cells in epithelium.

gonads **Organs** in animals which produce **gametes** and in some cases **hormones**. Examples are the **ovaries** and **testes**.

Graafian follicle A fluid-filled cavity in the mammalian **ovary** within which the **ovum** develops until **ovulation**.

greenhouse effect The increase in global temperature caused by increasing atmospheric carbon dioxide levels. This has resulted from burning **fossil fuels** and the destruction of large areas of tropical forest (so that less **photosynthesis** takes place). The increased carbon dioxide concentration traps the radiant **energy** of the Sun in a similar way to a greenhouse and may result in the polar ice-caps melting and a rise in sea level.

growth The increase in size and complexity of an organism during development from **embryo** to maturity, resulting from **cell division**, cell enlargement and

cell differentiation. In plants, growth originates at certain localized areas called *meristems*, while animal growth goes on all over the body.

guard cells Paired **cells** bordering **stomata** and controlling the opening and closing of the stomata. The **diffusion** of water into guard cells from adjacent **epidermis** cells causes the guard cells to expand and increase their **turgor**. However, they do not expand uniformly, the thicker, inelastic cell walls causing them to bend so that a pair of guard cells draws apart forming a stoma. Diffusion of water from guard cells reverses the process and closes the stoma. Stomata are normally open during the day and closed at night.

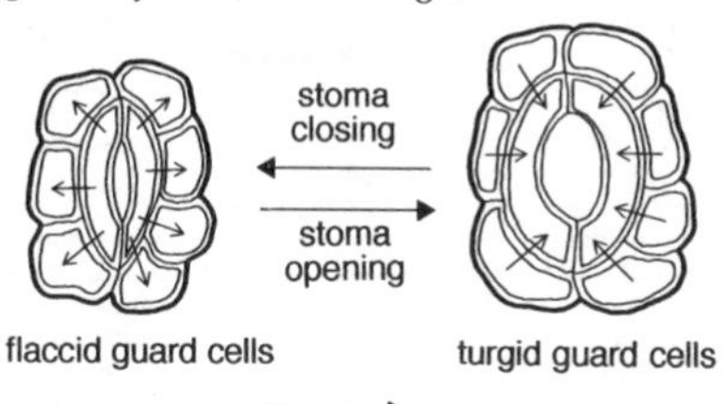

guard cells Stoma closing and opening.

gut All or part of the **alimentary canal**.

habitat The place where an animal or plant lives, the organism being adapted to the particular conditions within the habitat, which may be a woodland, sea-shore, pond, rockpool, etc. See **environment**.

haemoglobin Red pigment containing iron, within vertebrate **red blood cells**, responsible for the transport of oxygen throughout the body.

haemolysis The loss of **haemoglobin** from **red blood cells** as a result of damage to the **cell membrane**. This can be caused by several factors including **osmosis**, which can be investigated with human red blood cells which have a **solute** concentration equivalent to a 0.9% sodium chloride solution.

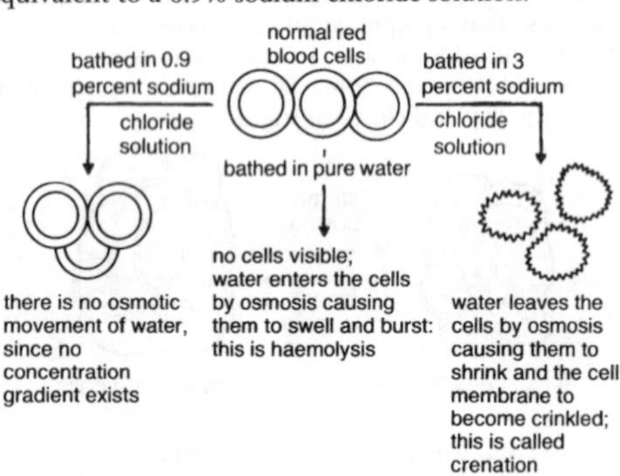

haemolysis Red blood cells in sodium chloride solution burst or shrink according to the concentration of the solution.

haemophilia See **sex linkage**.

haploid (used of a **nucleus, cell** or organism) Having a single set of unpaired **chromosomes**. The haploid number is found in plant and animal **gametes** as the result of **meiosis**. See **diploid**.

heart A muscular pumping organ which maintains **blood** circulation, and is usually equipped with **valves** to prevent backward flow. In mammals, the heart has four chambers, consisting of two relatively thin-walled *atria* (or *auricles*) which receive blood, and two thicker-walled *ventricles* which pump blood out.
The right side of the heart deals only with deoxygenated blood, and the left side only with oxygenated blood. The wall of the left ventricle is thicker and more powerful than that of the right, since it pumps all round the body, while the right ventricle pumps only to the lungs. See **circulatory system**, **heartbeat**.

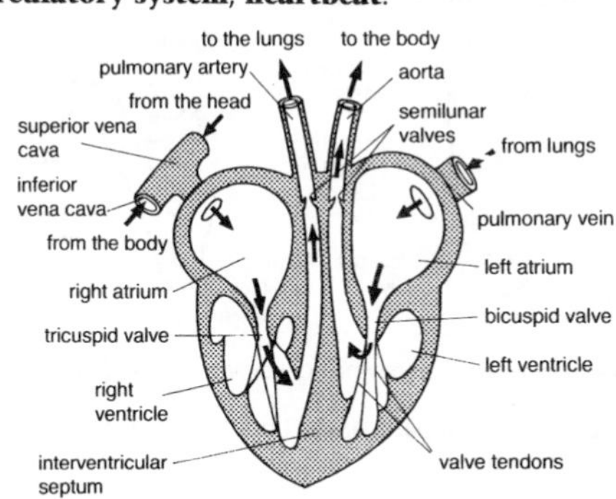

heart Structure of the mammalian heart.

heartbeat The alternate contraction and relaxation of

the **heart**. In mammals it consists of two phases
(a) *diastole*: the *atria* (singular, *atrium*) and **ventricles** relax, allowing **blood** to flow into the ventricles from the atria;
(b) *systole*: the ventricles contract, forcing blood into the **pulmonary artery** and **aorta**. The relaxed atria fill with blood in preparation for the next beat.

Heartbeat is initiated by a structure in the right atrium called the **pacemaker**, although the rate is controlled by the **medulla oblongata** of the **brain** which detects any increase in carbon dioxide in the blood as the result of increased **respiration** and is also affected by certain **hormones**, for example, **adrenalin** from the **adrenal gland**. The rate of human heartbeat is measured by counting the **pulse rate**.

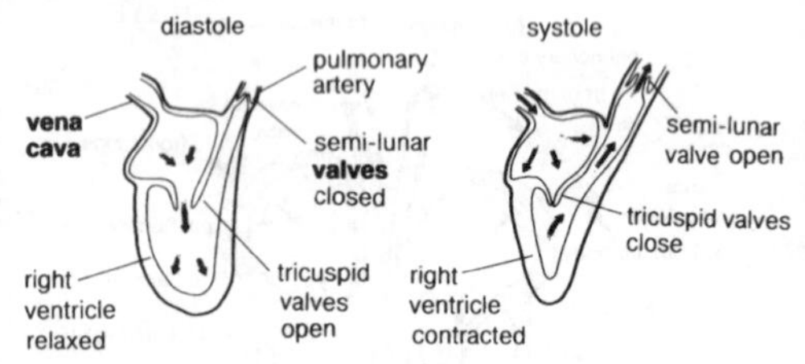

heartbeat Relaxation and contraction of heart (right side).

hepatic (used of parts of the body) Relating to the **liver** and its functions.

herbivore An animal which feeds on plants. Herbivores, which include sheep, rabbits and cattle, have a **dentition** adapted for chewing vegetation and a **gut**

capable of **cellulose digestion**.

Most herbivore **teeth** are grinders; **canines** are usually absent. In herbivores without upper **incisors**, a horny pad of gum combines with the lower incisors in biting vegetation. In most herbivores the lower jaw can move sideways, or backwards and forwards, thus producing the grinding action of the teeth.

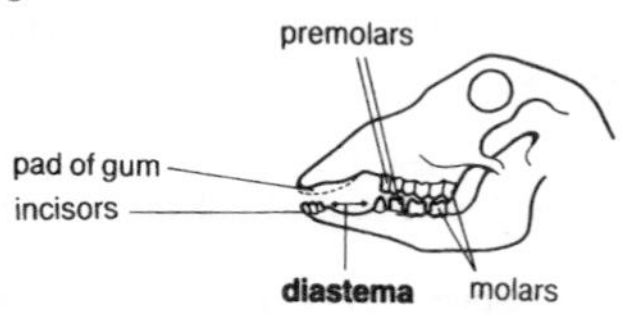

dental formula incisors 0/4 ; canines 0/0 premolars ; 3/3 molars 3/3 (total : 32)

herbivore The skull and teeth of a sheep.

heterotrophic (used of organisms) Able to obtain **organic compounds** (food) by feeding on other organisms. Heterotrophs include all animals and **fungi**, most **bacteria** and a few flowering plants. Heterotrophs are also called **food consumers** and can be classified into **carnivores, herbivores, omnivores, saprophytes and parasites**. Compare **autotrophic**.

heterozygous Having two different **alleles** of the same **gene**. See **monohybrid inheritance**.

homeostasis The maintenance of constant conditions within an organism. Examples are the control of **blood glucose** level by **insulin**, the control of blood water content by **ADH**, and the control of body temperature by the **skin**, etc.

homologous chromosomes Pairs of **chromosomes** that come together during **meiosis**. They carry **genes** that govern the same characteristics. Homologous pairs of chromosomes are found in all **diploid** organisms, one of the pair coming from the male **gamete** and the other from the female gamete, the pair being united at **fertilization**.

homozygous or **pure** Having two identical **alleles** for any one **gene**. See **monohybrid inheritance**.

hormones Chemicals secreted by the **endocrine glands** of animals and transported via the bloodstream to certain **organs** (*target organs*) where they cause specific effects which are vital in regulating and coordinating body activities. Hormone action is usually slower than nervous stimulation. The table summarizes the properties of some important human hormones; there are many others.

hormones (plant) See **plant growth substances**.

horticulture The science of plant growing which is of particular importance in food production.

human population curve A diagram showing changes in the human **population**. A population explosion has taken place in the last century due to improvements in food production and public health resulting in increasing **birth rate** and decreasing **death rate**. See diagram on page 115.

Endocrine gland	Hormone	Effects
Pituitary gland	ADH (anti-diuretic hormone)	Controls water reabsorption by the kidneys.
	TSH (thyroid stimulating hormone)	Stimulates thyroxine production in the thyroid gland.
	FSH (follicle-stimulating hormone)	Causes ova to mature and the ovaries to produce oestrogen.
	LH (luteinizing hormone)	Initiates ovulation and causes the ovaries to release progesterone.
	Growth hormone	Stimulates growth in young animals. In humans, deficiency causes dwarfism and excess causes gigantism.
Thyroid gland	Thyroxin	Controls rate of growth and development in young animals. In human infants, deficieny causes cretinism. Controls the rate of chemical activity in adults. Excess causes thinness and over-activity, and deficiency causes obesity and sluggishness.

Endocrine gland	Hormone	Effects
Pancreas (Islets of Langerhans	Insulin	Stimulates conversion of glucose to glycogen in the liver. Deficiency causes diabetes.
Adrenal glands	Adrenalin	Under conditions of 'fight, flight, or fright' causes changes which increase the efficiency of the animal. For example, increased heartbeat and breathing, diversion of blood from gut to muscles, conversion of glycogen in the liver to glucose.
Ovaries	Oestrogen	Stimulates secondary sexual characteristics in the female, for example, breast development. Causes the uterus wall to thicken during menstrual cycle.
	Progesterone	Prepares uterus for implantation.
Testes	Testosterone	Stimulates secondary sexual characteristics in the male, for example, facial hair.

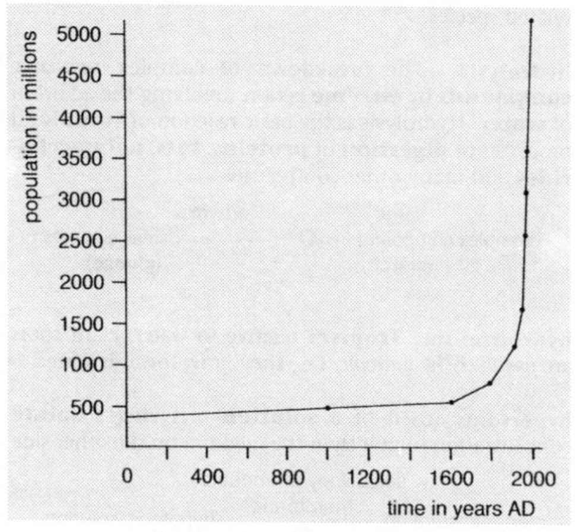

human population curve

humerus A **bone** of the upper forelimb of **tetrapods**. In humans, this is the bone of the upper arm. See **endoskeleton**.

humus The dark-coloured organic material in **soil** consisting of decomposing plants and animals, which provides nutrients for plants, and ultimately for animals. See **soil**.

hybrid A plant or animal produced as a result of a cross between two parents of the same **species** that are genetically unlike each other, or between two differing but related species.

hydrolysis The breakdown of complex **organic compounds** by **enzyme** action involving the addition of **water**. Hydrolysis is the basic reaction of virtually all processes of **digestion** of **proteins**, **fats**, **polysaccharides** and many other compounds.

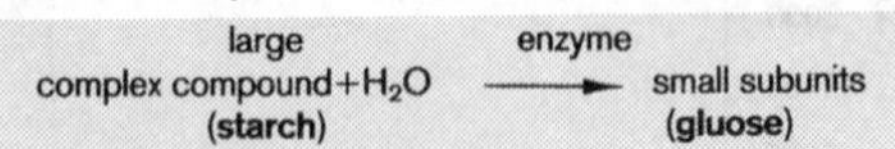

hydrotropism **Tropism** relative to water. Plant roots are *positively hydrotropic*, i.e., they grow towards water.

hypertonic (used of a **solution**) Having a **solute** concentration higher than the medium on the other side

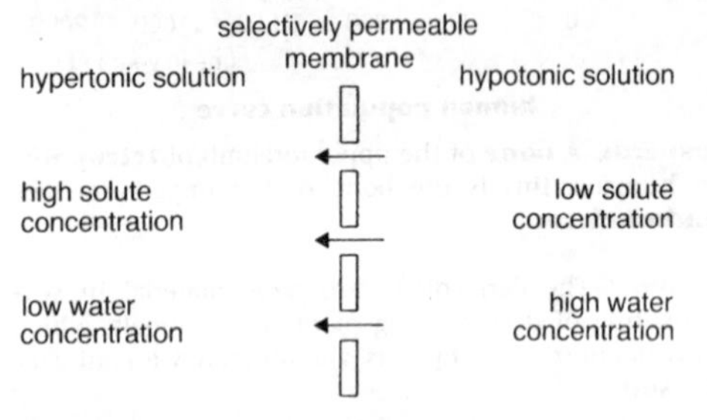

hypertonic

of a **selectively permeable membrane**. Such a solution will gain water by **osmosis**. See **hypotonic**, **isotonic**. See diagram.

hypothesis A suggested solution to a scientific problem which must be tested by experimentation and, if not validated, must then be discarded. See **scientific method**.

hypotonic (used of a **solution**) Having a **solute** concentration lower than the medium on the other side of a **selectively permeable membrane**. Such a solution will lose water by **osmosis**. See **hypertonic**, **isotonic**.

ileum The final region of the mammalian **small intestine** which receives food from the **duodenum**. The lining of the ileum secretes **enzymes** which complete the **digestion** of **protein**, **carbohydrate** and **fat**, into **amino acids**, simple sugars (mainly **glucose**), fatty acids, and glycerol.
Absorption of food occurs in the ileum which has a large absorbing surface consisting of thousands of finger-like structures called **villi**. The lining of each villus is very thin, allowing the passage of soluble foods, and each contains a network of **blood capillaries**.
Amino acid and glucose particles diffuse into the blood capillaries from where they are transported first to the **liver** and then to the general circulation. Fatty acid and glycerol particles pass into the **lacteals** and are circulated via the **lymphatic system**.
Material not absorbed, for example, **roughage**, is passed into the **large intestine**. See **digestion, assimilation**. See diagram overleaf.

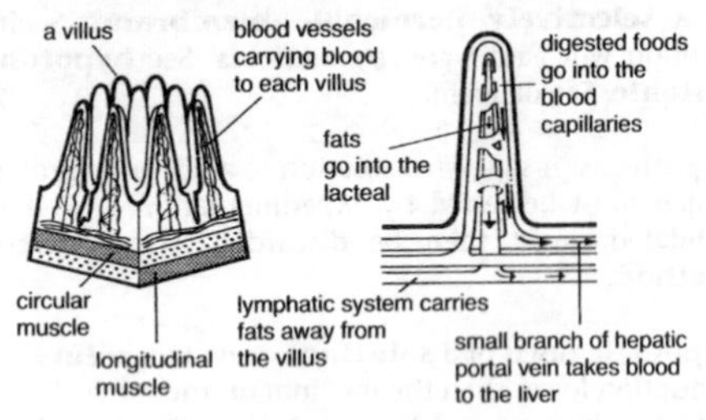

ileum Section through the ileum wall (left) and a section of a villus (right).

imago An adult, sexually mature insect. See **metamorphosis**.

immobilization The process whereby the movement of **cells** or **enzymes** is restricted by attachment to another substance. For example, enzymes can be fixed onto glass beads in such a way that, having catalysed the required chemical reaction, they can be easily separated from the product and re-used. Immobilization techniques are important in **biotechnology**.

glass bead + enzymes → immobilized enzymes on bead

immobilization

immunity The ability of the body to resist infectious diseases. Once the body has suffered from a disease and produced specific **antibodies**, these can remain in the **blood** for months or even years after they helped the organism to recover from the disease. While they are there, they stop the organism catching the same disease again. Those who have antibodies against a disease, or can make them, are *immune* from that disease.

There are different types of immunity:

active immunity results when an organism makes its own antibodies as a result of contact with an **antigen**. The antigen may be a **pathogen** or its product. Active immunity usually develops slowly, yet it may last for many years.

Active immunity may be induced artificially by the introduction of antigens into the body. Such antigens may be composed of living, dead or weakened pathogens which are used to stimulate antibody production but not cause the disease. Suspensions of these pathogens are called **vaccines**.

Passive immunity is given when antibodies produced by active immunity in one organism are transferred to another. The antibodies are extracted from the **blood serum** of a person or animal (usually a horse) which has recovered from the disease. Since no antigen is introduced, there is no stimulation of antibody production; hence passive immunity does not last long, but it does provide immediate protection, unlike active immunity which requires time for development. The introduction of existing antibodies is done if the body is infected with a disease that is too dangerous to leave until the body's natural defences are activated, or as a preventative measure to guard against such an infection. See

immunization.

immunization The prevention of infection by artificially introducing a **pathogen** into the body. This stimulates **antibody** production and causes the organism to become immune to the pathogen, which in some cases lasts a lifetime. See **immunity**. In Britain, children are immunized against polio, diphtheria, tetanus, whooping cough and tuberculosis. Girls are also immunized against German measles, a disease which can harm an unborn child. People going abroad may be immunized against diseases such as yellow fever.

implantation Attachment of a mammalian **embryo** to the **uterus** lining at the start of **pregnancy**. In preparation for implantation, the uterus wall becomes thicker with new **cells** and an increased **blood** supply. See **fertilization**.

incisors Chisel-shaped cutting **teeth** at the front of the mouth used for biting off pieces of food. See **dental formula**, **dentition**, **carnivore**, **herbivore**, **omnivore**.

incomplete dominance A genetic condition in which neither of a pair of **alleles** is dominant but instead they 'blend' to produce an intermediate trait. See **monohybrid inheritance**.

incubator **1.** A heated container used to grow **microorganisms** on **nutrient agar plates**. The usual incubation period is 48 hours at 37 °C.
2. A heated chamber used to maintain the body temperature of premature babies.

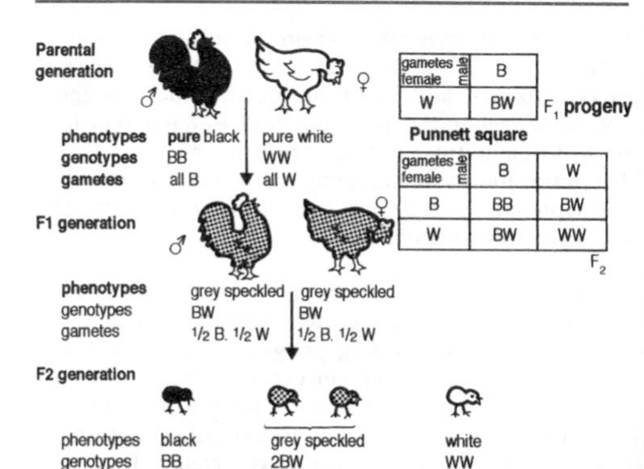

incomplete dominance The inheritance of feather colour in Andalusian fowls.

indicator organism An organism which can survive only in certain environmental conditions, and hence one whose presence provides information about the **environment** in which it is found. For example, the **bacterium** *Escherichia coli* lives in animal **gut** and is always present in **faeces**. Although *E. coli* is itself harmless, its presence in water indicates sewage **pollution**.

inherited diseases Disorders which are hereditary, i.e. they are passed from generation to generation by the **genes**. Examples are **haemophilia**, sickle-cell anaemia, cystic fibrosis and Huntington's chorea.
Inherited disorders are due to rare **genotypes** producing disease-susceptible **phenotypes**.

inorganic compounds Chemical substances within **cells** which are derived from the external physical **environment**, and which are not organic. The most abundant cell inorganic compound is water which is present in amounts ranging from 5 to 90%.
The other inorganic components of cells are **mineral salts** present in amounts ranging from 1 to 5%. See **organic compounds**.

insulin A vertebrate **hormone** secreted by the *islets of Langerhans* in the **pancreas**. Insulin regulates the conversion of **glucose** to **glycogen** in the **liver**. If the concentration of **blood** glucose is high, the rate of secretion of insulin is high, and thus glucose is rapidly converted to liver glycogen. If the concentration of blood glucose is low, less insulin is secreted. This is an example of *feedback regulation* found in relation to many hormones. People who suffer from diabetes produce insufficient insulin to control the delicate glucose level balance in their bodies.

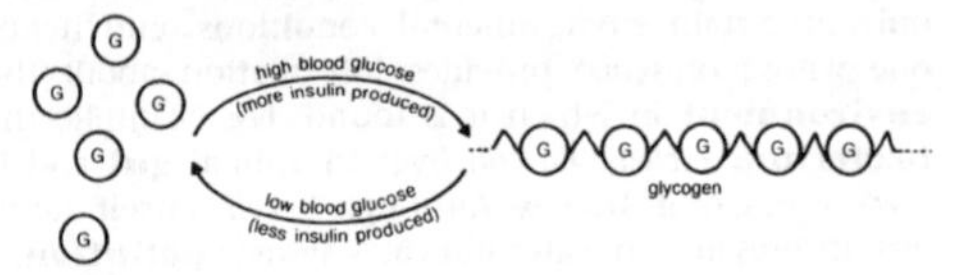

insulin Feedback regulation of insulin secretion.

integrated control The control of **pests** using a combination of chemical and **biological control**.

integument **1.** The external protective covering of an animal, e.g. **skin**, **cuticle**.

2. The protective layer around flowering plant ovules which, after **fertilization**, forms the **testa**.

intercostal muscles The **muscles** positioned between the ribs of mammals, that are important in **lung** ventilation. See **breathing**.

intestine The region of the **alimentary canal** between the **stomach** and the **anus** or **cloaca**. In vertebrates it is the major area of **digestion** and **absorption** of food, and is usually differentiated into an anterior **small intestine** and a posterior **large intestine**. See **digestion**.

in vitro (used of biological experiments or observations) Conducted outside an organism, for example in a test tube.

in vivo (used of biological experiments or observations) Conducted within living organisms.

involuntary muscle or **smooth muscle** The type of **muscle** associated with internal **tissues** and **organs** in mammals, for example, the **gut** and **blood vessels**. It is called involuntary muscle because it is not directly controlled by the will of the organism. Involuntary muscle actions include contraction and dilation of the **pupil** in the **eye**, and **peristalsis**. See **voluntary muscle, antagonistic muscles**.

ion An electrically charged **atom** or group of atoms.

iris The structure in the vertebrate **eye** which controls the size of the **pupil** and hence the amount of light entering the **eye**.

irritability See **sensitivity**.

isotonic (used of a **solution**) Having a **solute** concentration equal to that of the medium on the other side of a **selectively permeable membrane**. There is no water movement across the membrane by **osmosis.** See **hypertonic**, **hypotonic**.

joint The point in a **skeleton** where two or more **bones** meet and movement may be possible. Moveable joints in mammals are of three types:
(a) *ball and socket joints* which allow movement in several planes.

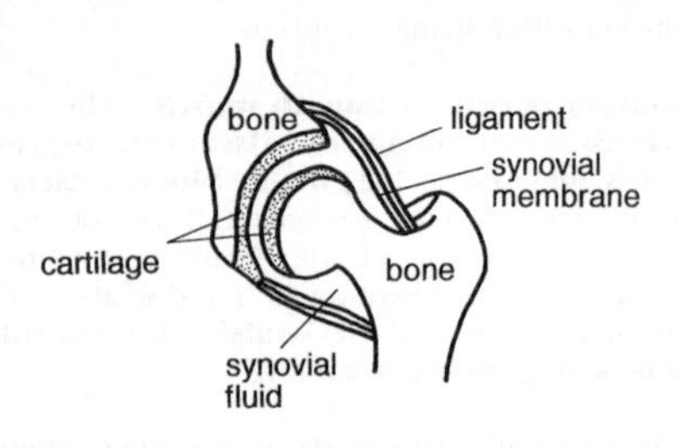

joint Ball and socket joint.

(b) *hinge joints* which allow movement in only one plane.

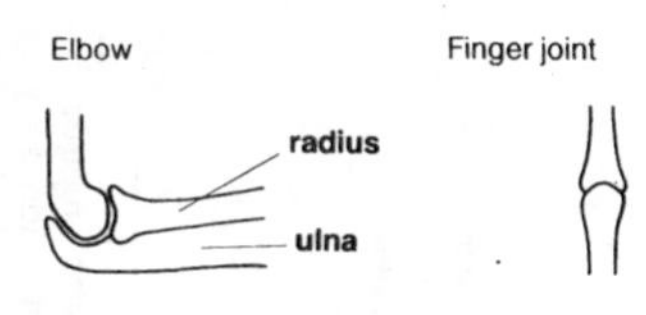

joint Hinge joint.

(c) *gliding joints* which occur when two flat surfaces glide over one another, allowing a small amount of movement only.

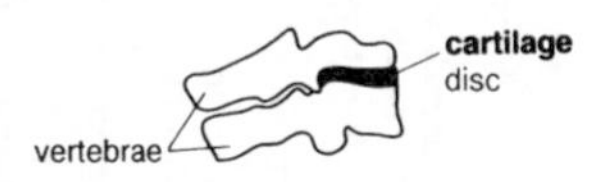

joint Gliding joint.

keratin Strong, fibrous **protein** present in vertebrate **epidermis** forming the outer protective layer of **skin** and also hair, nails, wool, feathers, and horns.

key A device for identifying unfamiliar organisms. The two types of key shown can be used to identify the major divisions of living organisms:

(a) *branching key*: start at the top and follow the branches by deciding which descriptions best fit the organism to be identified;

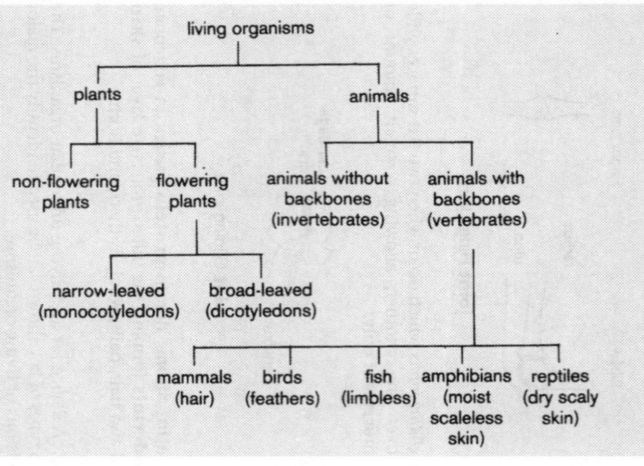
living organisms
plants
animals
non-flowering plants
flowering plants
animals without backbones (invertebrates)
animals with backbones (vertebrates)
narrow-leaved (monocotyledons)
broad-leaved (dicotyledons)
mammals (hair)
birds (feathers)
fish (limbless)
amphibians (moist scaleless skin)
reptiles (dry scaly skin)

key

(b) *numbered key*: at each stage, there is a pair of alternative statements, only one of which can apply to the specimen to be identified. The correct alternative leads to the next choice and so on until the organism is named.

This type of key is preferable to the branching type which can take up too much space.

1	a	Multicellular, usually with chlorophyll	Go to 2
	b	Multicellular, without chlorophyll, able to move	Go to 3
2	a	Plants which reproduce using flowers	Go to 4
	b	Plants which reproduce without flowers	Non-flowering plants
3	a	Animals with backbones	Go to 5
	b	Animals without backbones	Invertebrates
4	a	Narrow-leaved plants with one cotyledon in their seeds	Monocotyledons
	b	Broad-leaved plants with two cotyledons in their seeds	Dicotyledons
5	a	Hair present	Mammals
	b	Hair absent	Go to 6
6	a	Feathers present	Birds
	b	Feathers absent	Go to 7
7	a	Dry, scaly skin	Reptiles
	b	Other type of skin	Go to 8
8	a	Four limbs, moist scaleless skin	Amphibians
	b	Limbless	Fish

Using the above key, it is now possible to do a simple classification of organisms such as ape, pike, newt, oak, snake, grass and earthworm.

kidney The **organ** of **excretion** and **osmoregulation** in vertebrates, consisting of units called **nephrons**. In humans, the kidneys are a pair of red-brown oval structures at the back of the **abdomen**.

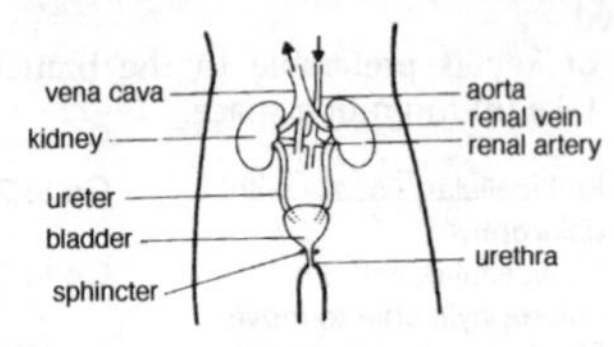

kidney The position of the kidneys and associated organs in the human body.

Oxygenated **blood** enters each kidney via the renal **artery**, and the renal **vein** removes deoxygenated blood. Another tube, the **ureter**, connects each kidney with the **bladder**.

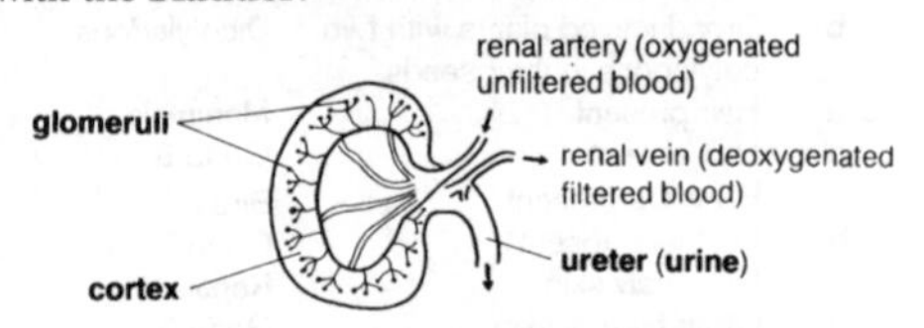

kidney Section through a kidney.

The renal artery divides into numerous *arterioles* which terminate in tiny knots of blood **capillaries** called **glomeruli**. Each glomerulus (there are about one million in a human kidney) is enclosed in a cup-shaped organ called a **Bowman's capsule**.

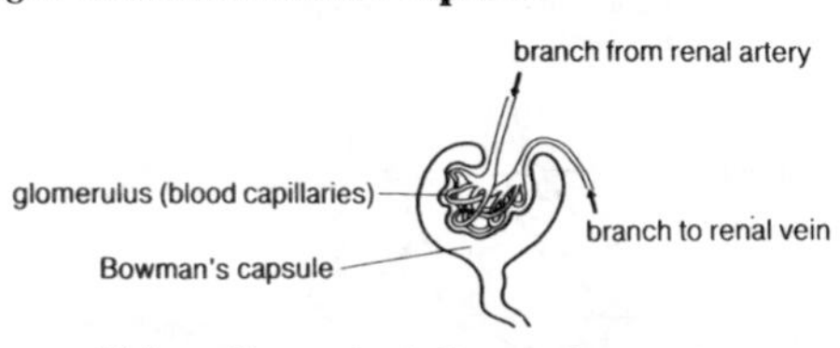

kidney Glomerulus in Bowman's capsule.

Two processes occur in the kidneys:

(a) *ultrafiltration*. the vessel leaving each glomerulus is narrower than the vessel entering, causing the blood in the glomerulus to be under high pressure, which forces the blood components with smaller molecules through the **selectively permeable** capillary wall into the Bowman's capsule.

Large particles (unfiltered)	Small particles (filtered)
Blood cells	**Glucose**
Plasma proteins	**Urea**
	Mineral salts
	Water
	Amino acids

(b) *reabsorption*: the fluid filtered from the blood (*filtrate*) passes from the Bowman's capsule down the renal tubule where reabsorption of useful materials occurs, i.e all the glucose and amino acids, and some of the salts and water are reabsorbed into the blood.

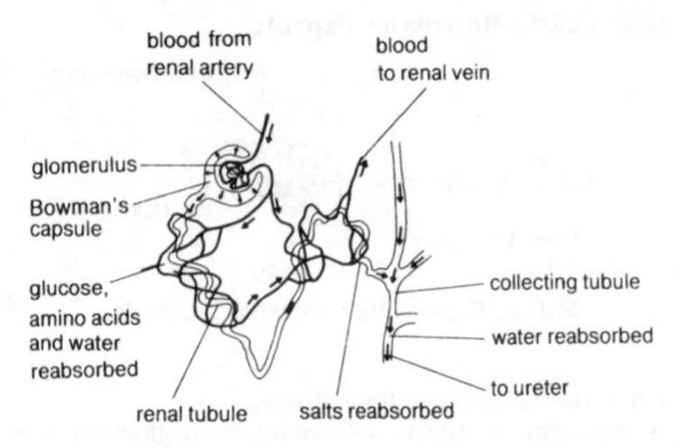

kidney Reabsorption of useful substances.

The liquid remaining after filtration and reabsorption by the kidney is a solution of salts and urea in water (*urine*) and is passed to the **bladder** via the ureters from where it is expelled via the **urethra** under the control of a **sphincter muscle**. See **antidiuretic hormone**.

kidney machine An artificial **kidney** which purifies the **blood** of a person with diseased kidneys. Blood is taken from the person's arm artery and passed through a **selectively permeable membrane** where **urea** and excess **mineral salts** are removed into a rinsing solution. This process is called *dialysis*.
The purified blood is then returned to the person's arm **vein**.

kilojoule The unit used to measure the energy content of food. See **food calorimeter**.

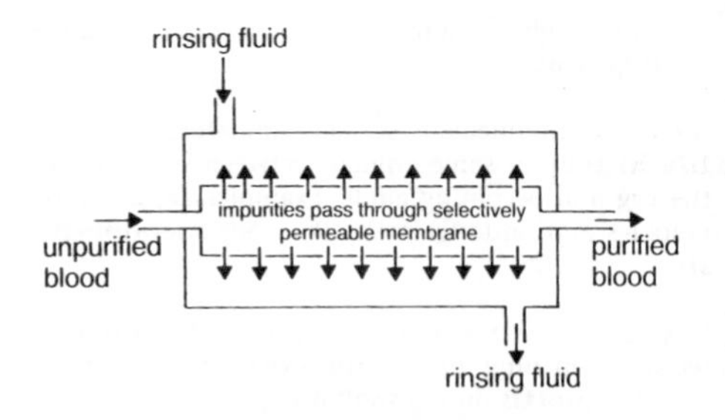

kidney machine

kingdom Any of the five great divisions of living organisms, i.e. the Animal, Plant, **Protista**, **Fungi** and **Bacteria** kingdoms. See **classification**.

lacteals **Lymph** vessels within the **villi** of the vertebrate **intestines**. The products of the **digestion** of **fats** (fatty acids and glycerol) diffuse into the lacteals and are circulated via the **lymphatic system**.

lactic acid An organic acid (CH_3 CH OH COOH) produced during **respiration** in many animal **cells**, including vertebrate **muscle** cells, and certain **bacteria**. See **oxygen debt**.

large intestine The posterior region of the vertebrate **intestine**. In humans, at the entry to the large intestine there is a region called the **caecum**, from which projects the **appendix**, but most of the large intestine consists of the **colon** which leads to the **rectum**. The large intestine receives undigested material from the **ileum**. See **digestion**.

larva An intermediate, sexually immature stage in the **life history** of some animals between hatching from the egg and becoming adult. Examples are amphibian tadpoles and butterfly caterpillars. See **metamorphosis**.

larynx A region at the upper end of the **trachea** of tetrapods opening into the **pharynx** and specialized to close the **glottis** during swallowing.
In mammals, amphibians, and reptiles, vocal cords within the larynx produce sound. See **breathing**.

leaching The removal of **mineral salts** and other dissolved substances from the **soil** by the downward movement of **water**, mainly rain. The substances are carried into streams and rivers.

leaf That part of a flowering plant which grows from the **stem** and is typically flat and green. The functions of leaves are (a) **photosynthesis**, (b) **gas exchange**, and (c) **transpiration**. See **adaptation** for a detailed description of leaf structures.

lens A transparent structure behind the **pupil** of the vertebrate **eye**, important in focusing the image on the **retina**, and in **accommodation**.

lenticel One of many pores developing in woody **stems** and **roots** when **epidermis** is replaced by bark, through which **gas exchange** occurs.

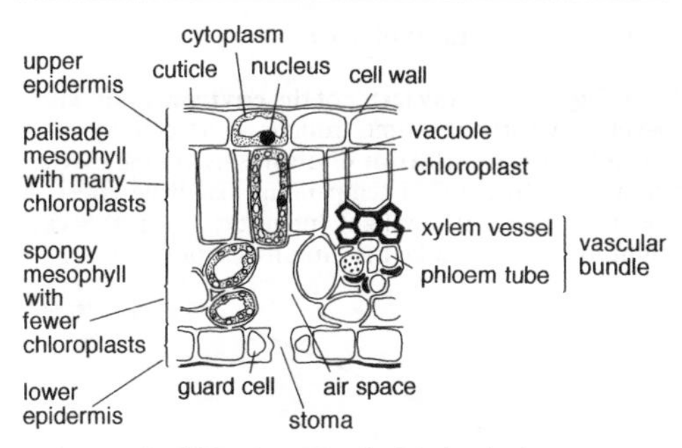

leaf Structure of a dicotyledon leaf.

leucocyte See **white blood cell**.

lichen A plant formed by **mutualism** between an **alga** and a **fungus**. The alga supplies **carbohydrate** and oxygen to the fungus, and receives **water** and **mineral salts** in return.

life history or **life cycle** The various stages of development which organisms undergo from egg to adult. See **metamorphosis**.

ligament A strong band of **collagen** connecting the **bones** at moveable vertebrate **joints**. Ligaments strengthen the joint, allowing movement in only certain directions and preventing dislocation.

lignin An **organic compound** deposited in the **cell** walls of **xylem** vessels, giving strength. Lignin is an important constituent of *wood*.

limiting factor Any factor of the **environment** whose level at a particular time inhibits some activity of an organism or **population** of organisms. As the diagram shows, an Increase in temperature has little effect on photosynthesis at low light intensities, so in this case, light intensity must be the limiting factor.

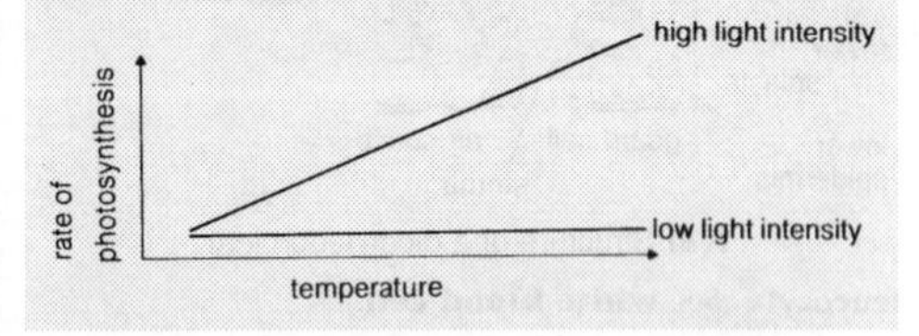

limiting factor Photosynthesis is limited by low light intensity even if temperature increases.

lipase An **enzyme** which digests **fat** into fatty acids and glycerol by **hydrolysis**. In mammals, lipase is secreted by the **pancreas** and the **ileum**. See diagram.

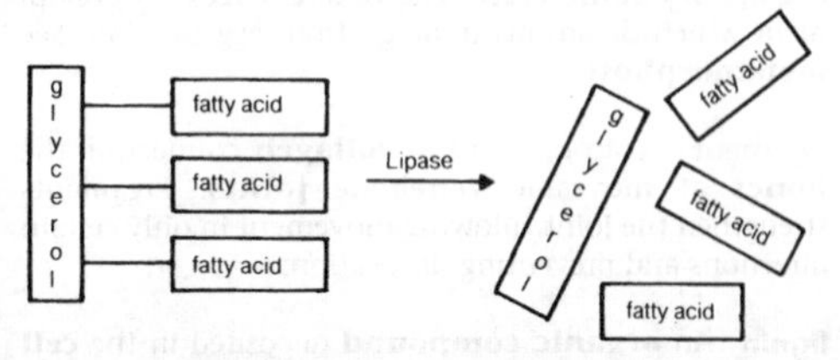

lipase Hydrolytic breakdown of fat.

lipid See **fat**.

liver The largest organ of the vertebrate body, occupying much of the upper part of the **abdomen**, in close association with the **alimentary canal**. See **digestion**, **circulatory systems**.
Some of the many functions of the liver are
(a) production of **bile**;
(b) **deamination** of surplus **amino acids**;
(c) regulation of **blood** sugar by interconversion of **glucose** and **glycogen**;
(d) storage of iron, and **vitamins** A and D;
(e) detoxication of poisonous by-products;
(f) release and distribution of heat produced by the chemical activity of liver cells;
(g) conversion of stored fat for use by the **tissues**;
(h) manufacture of **fibrinogen**.

long sight or **hypermetropia** A human **eye** defect mainly caused by the distance from **lens** to **retina** being shorter than normal. This results in near objects being focused behind the retina giving blurred vision. Long sight is corrected by wearing converging (*convex*)

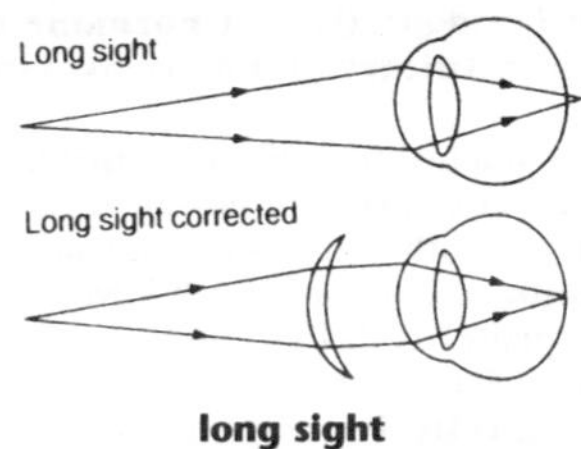

long sight

lenses.

lungs The **organs** of **breathing** in mammals, amphibians, reptiles, and birds. In mammals, the lungs are two elastic sacs in the **thorax** which can be expanded or compressed by movements of the thorax in such a way that air is continually taken in and expelled. The **trachea** (windpipe) connects the lungs with the atmosphere. It divides into two **bronchi** which enter the lungs and further divide into many smaller *bronchioles* which terminate into millions of air-sacs called **alveoli** which are the **gas exchange** surface and which are in close contact with **blood vessels** bringing blood from and to the **heart**.

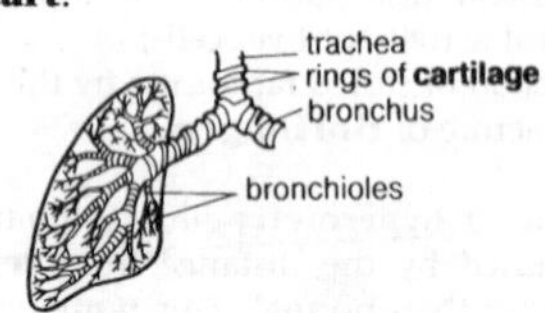

lungs Air passages in the lung.

luteinizing hormone (LH) A **hormone** secreted by the vertebrate **pituitary gland**. See **ovulation**.

lymph Fluid drained from **blood capillaries** in vertebrates as a result of high pressure at the arterial end of the capillary network. Lymph or *tissue fluid* which is similar to **plasma** (except for a much lower concentration of **plasma proteins**) bathes the **tissues** and acts as a medium in which substances are exchanged between capillaries and **cells**. For example, oxygen and glucose diffuse into the cells while carbon dioxide and **urea** are removed. Lymph drains back into capillaries or into vessels called lymphatics which then connect with the general circulation via the **lymphatic system**.

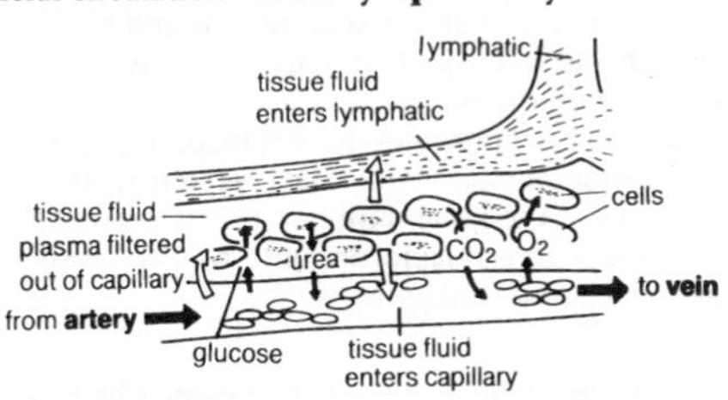

lymph The lymphatics, capillaries and cells.

lymphatic system A system of fluid-containing vessels (*lymphatics*) in vertebrates, which return lymph to the general **blood** circulation. The lymphatic system is also important in
(a) transporting the products of **fat digestion**;
(b) production of **white blood cells** and **antibodies**.

lymph nodes Structures within the **lymphatic system** which filter **bacteria** from **lymph** and produce **white blood cells** and **antibodies**.

lymphocyte A type of **white blood cell** of vertebrates. There are two types of lymphocytes:
B-lymphocytes which are derived from the bone marrow. These cells make and release **antibodies** into the bloodstream to attack **antigens**.
T-lymphocytes which are derived from the thymus gland.

These make antibodies which remain attached to the surface of the cell. It is therefore the lymphocyte itself which attacks the antigens. There are several sorts of T-lymphocyte. Some of them destroy cancer cells; others recognize and attack infected body cells and foreign tissue which has been grafted into the body. These are called the *killer T-cells*.
Some B- and T-lymphocytes, having made a specific antibody, become *memory cells*. These remain in the body ready to make the same antibody if it is needed again. This provides a very rapid response to any subsequent invasion by similar antigens.

malnutrition Inadequate nutrition caused by lack of a **balanced diet** or by insufficient food. In tropical countries, diseases associated with **protein** and **carbohydrate** deficiency cause many deaths among young children.

medulla The central part of a **tissue** or **organ** such as the mammalian **kidney**. See **cortex**.

medulla oblongata The posterior region of the vertebrate **brain** which is continuous with the **spinal cord** and which in mammals controls **heartbeat**, **breathing**, **peristalsis** and other involuntary actions.

meiosis A method of *nuclear division* that occurs during the formation of **gametes** when a **diploid nucleus** gives rise to four **haploid** nuclei. There are two consecutive divisions in the process, as shown below.

Interphase
contents of nucleus indistinct

Prophase
contents of nucleus become clear

Homologous chromosomes form pair. Nuclear membrane breaks down

each chromosome can be seen to consist of two chromatids joined by a centromere

Metaphase
nuclear spindle forms and chromosomes move to the equator

Anaphase
centromeres repel each other carrying chromosomes towards the poles of the spindle

Telophase
nuclear membranes form around the groups of chromosomes

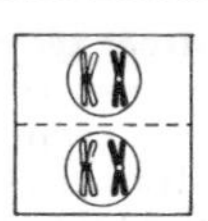

meiosis Stages in the two divisions of the process.

During the first division of meiosis the **homologous chromosomes** may exchange genetic material as they lie side by side, and this leads to **variation** in the resulting nuclei. This process is called *crossing over*.

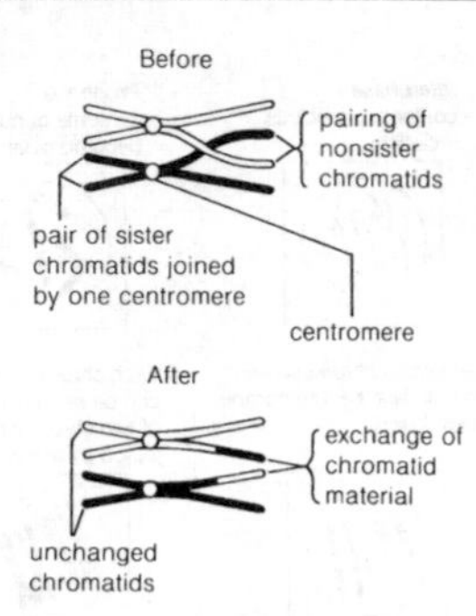

meiosis Crossing over between non-sister chromatids.

menopause The time at the end of the reproductive period of human females when **menstrual cycles** cease to occur.

menstrual cycle The reproductive cycle occurring in female primates (monkeys, apes, humans). This cycle is under the control of **hormones**. In human females, the cycle lasts about 28 days, during which the **uterus** is prepared for **implantation**. If **fertilization** does not occur, the new uterus lining and unfertilized **ovum** are expelled, which results in bleeding from the **vagina (menstruation)**. See **ovulation**.

menstruation The discharge of **uterus** lining-**tissue** and **blood** from the **vagina** in human females at the end of a **menstrual cycle** in which **fertilization** has not occurred.

messenger RNA See **RNA, protein synthesis.**

metabolic water One of the products of aerobic respiration which is an important source of water for desert animals.

metabolism The sum of all the physical and chemical processes occurring within a living organism. These include both the synthesis (*anabolism*) and breakdown (*catabolism*) of compounds. See **basal metabolic rate**

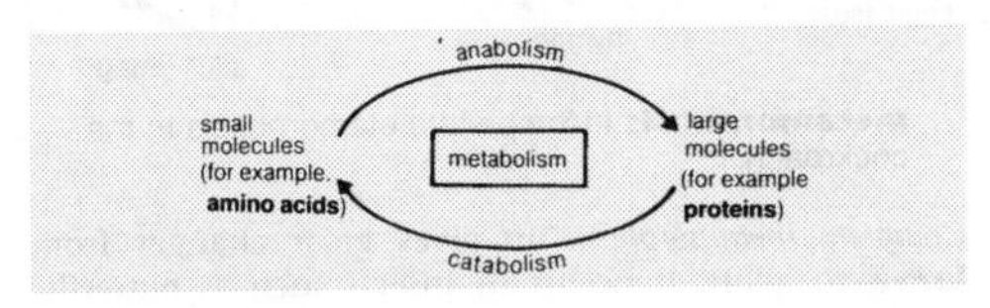

metabolism The synthesis and breakdown of compounds within an organism.

metamorphosis The period in the **life history** of some animals when the juvenile stage is transformed into an adult.

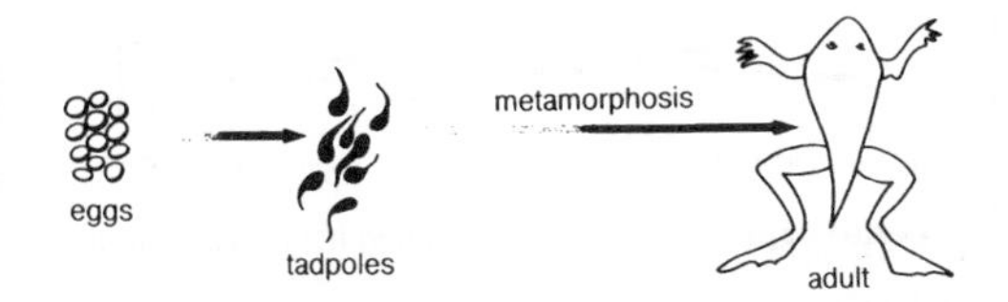

metamorphosis Amphibians go through a tadpole stage.

Incomplete metamorphosis is a type of development in which there are relatively few changes from juvenile form to adult. It occurs in insects such as the dragonfly, locust and cockroach, in which the juvenile form (**nymph**) resembles the adult except that it is smaller, wingless, and sexually immature.

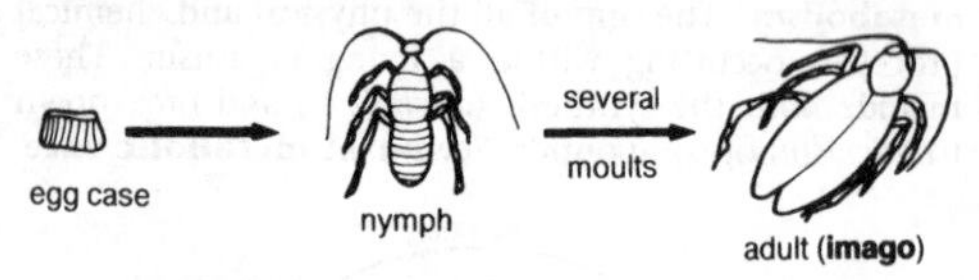

metamorphosis Incomplete metamorphosis in the cockroach.

Complete metamorphosis involves great changes from **larva** to adult. It occurs in insects such as butterfly, moth, housefly, etc., and the larvae in such life histories are maggots, grubs, or caterpillars (depending on the species) and are quite unlike the adult form. A series of moults (**ecdyses**) produces the **pupa** which then becomes completely reorganized and develops into the adult, the only sexually mature stage.

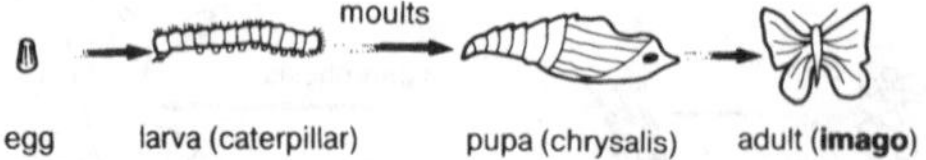

metamorphosis Complete metamorphosis in a butterfly.

microbe An alternative term for **microorganism**.

microbiology The study of **microorganisms**.

microbiology experiments The use of sterile *nutrient*

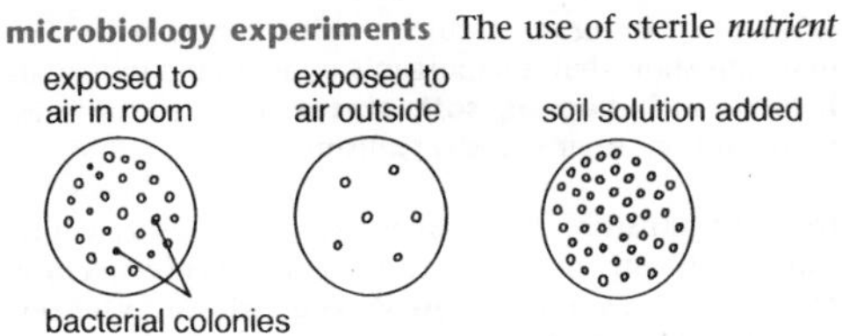

river water added

plate coughed on

rubbed with sterile swab

rubbed with swab which
has been rubbed on floor

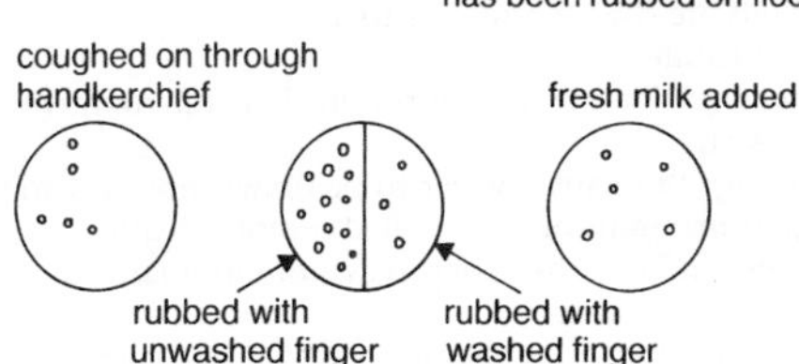

sour milk added

fresh cheese

stale cheese

moulds

microbiology experiments

agar plates (see **agar**) to culture **microorganisms**. The diagrams show that microorganisms are found in various habitats including air, **soil**, **water**, exposed surfaces, food, and on other living organisms.

microorganisms Very small living organisms which can usually only be seen with the aid of a **microscope**. Microorganisms include **protozoa**, **algae**, **viruses**, **fungi** and **bacteria**.

microorganisms and food Most foods contain **microorganisms** which are usually destroyed during cooking. Some foods must be treated before consumption, an example being milk, which is heated at 80 °C for 30 seconds. This is called *pasteurization*.
Microorganisms cause food to go off, and various methods are used to preserve food.
These include:
(a) *freezing*: low temperatures inhibit microorganism growth;
(b) *salting*: this causes water to be drawn from microorganisms resulting in so-called osmotic death;
(c) *dehydration*: this deprives microorganisms of necessary water.
(d) *smoking*: this adds chemicals which kill microorganisms;
(e) *high temperature sterilization*: this is used in the canning of foods.
(f) *chemical preservatives* e.g. an **acid** such as acetic acid (vinegar) which is added to some foods to cause an acid **pH** unfavourable to microorganisms.

micropyle **1.** A pore in a **seed** through which water is

absorbed at the start of **germination**
2. A pore in the **ovule** of a **flower** through which the **pollen** tube delivers the male **gamete**.
3. A pore in the **ovum** of insects through which the **spermatozoon** enters.

microscope An instrument used to magnify structures, for example, **cells** or organisms, which are not visible to the naked eye:

(a) *light microscope*: light illuminates the specimen which is magnified by glass lenses.
The magnification of the microscope is found by multiplying the magnification of the objective lens (e.g. x 40) by the magnification of the eyepiece lens (e.g. x 10) to give the total magnification (in this example x 400). The maximum possible magnification using a light microscope is x 1500. *Thin specimens* are placed on a glass slide and may be stained with dyes which show up particular structures;
(b) *phase-contrast microscope*: this allows the viewing of transparent and unstained structures;
(c) *electron microscope*: the most advanced type, giving magnification as high as x 500 000.

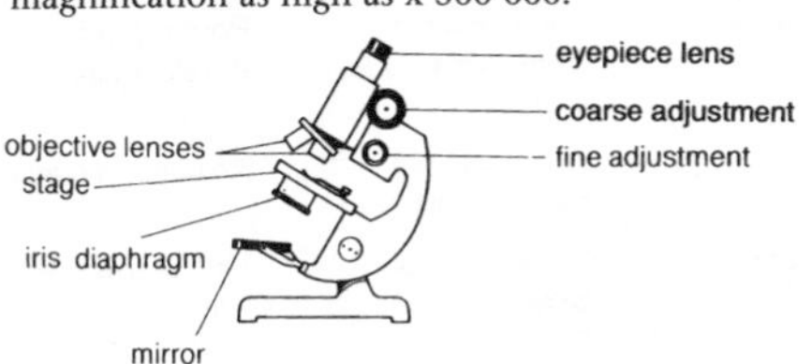

microscope Parts of a light microscope.

milk teeth or **deciduous teeth** The first set of **teeth** occurring in most mammals. For example, humans have 20 milk teeth which are replaced during childhood by the larger permanent teeth.

mineral salts Components of **soil** formed from rock **weathering** and **humus** mineralization, and found in **solution** in soil water. Mineral salts are absorbed by plant **roots** and transported through the plant in the **transpiration stream**. Like **vitamins**, mineral salts are required in tiny amounts, but are nevertheless vital for plant and ultimately animal nutrition; the absence of

Mineral salt	Function	Some effects of deficiency
Phosphorus	Components of ATP, nucleic acids, cell membrane, animal bones.	Stunted plant growth.
Calcium	Component of plant cell walls and animal bones.	Rickets in humans.
Nitrogen	Component of protein and nucleic acids.	Poor reproductive development in plants.
Iron	Component of haemoglobin.	Anaemia in humans.
Magnesium	Component of chlorophyll.	Pale yellow plant leaves (chlorosis).

mineral salts Some important minerals and their functions.

a particular mineral salt can lead to mineral deficiency disease and death. Plants require at least twelve mineral salts for healthy growth:

(a) *essential elements* required in relatively large quantities: nitrogen, phosphorus, sulphur, potassium, calcium, magnesium;

(b) *trace elements* required in very small amounts: manganese, copper, zinc, iron, boron, molybdenum.

Some mineral salts are required by plants, some are required by animals, and some by both.

mitochondrion A microscopic **cell organelle** in the **cytoplasm** of aerobic cells, in which some **respiration** reactions occur. The inner membrane of a mitochondrion wall is highly folded, giving rise to a series of partitions called *cristae*, which greatly increase the surface area for the attachment of respiratory enzymes. See **aerobe, enzymes**.

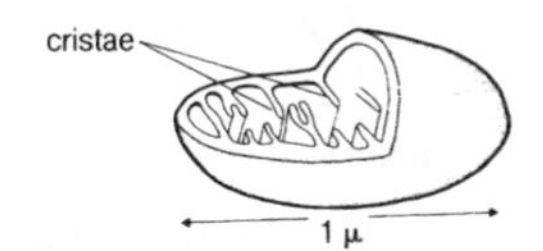

mitochondrion Section showing cristae.

mitosis The process necessary for the growth of an organism, by which the **nucleus** of a **cell** divides in such a way that the resultant *daughter cells* receive precisely the same numbers and types of **chromosomes** as the original *mother cell*. During this type of nuclear division the chromosomes of the mother cell (the dividing cell) are first duplicated and then passed in identical

sets to the two daughter cells.
In stage 1, Chromosomes appear as coiled threads. Two structures called *centrioles* are seen outside the nucleus.

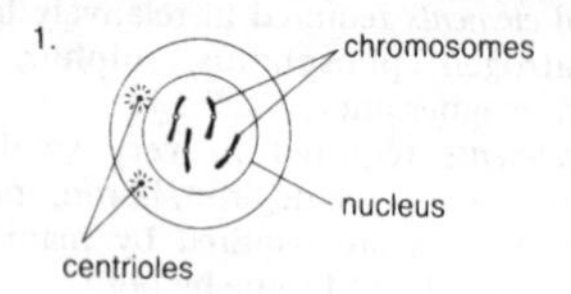

mitosis Stage 1.

Each chromosome makes a duplicate of itself. These duplicates or *chromatids* are joined at a *centromere*.

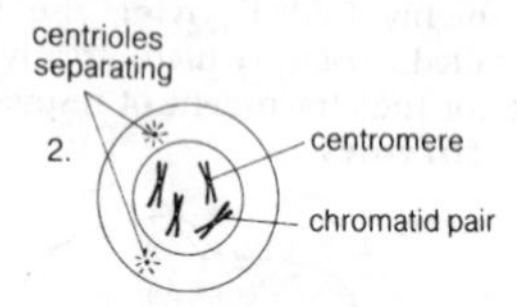

mitosis Stage 2.

The *nuclear membrane* disappears. The centrioles move to opposite poles of the cell and produce a system of fibres called a *spindle*. The chromatid pairs line up at the equator of the cell.

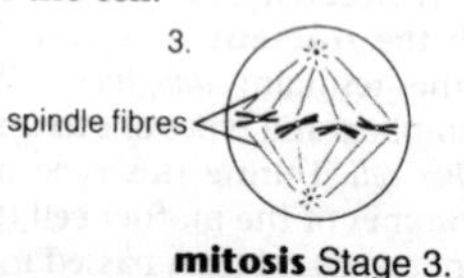

mitosis Stage 3.

The chromatid pairs separate and move to opposite poles of the cell.

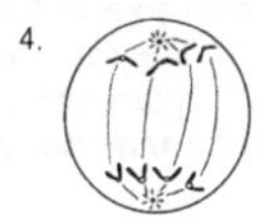

mitosis Stage 4.

New nuclear membranes form and the **cytoplasm** starts to divide.

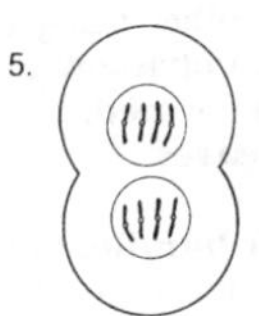

mitosis Stage 5.

The chromosomes disappear and the cells return to the resting state.

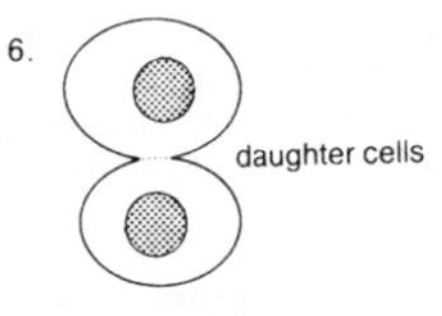

mitosis Stage 6.

molars Broad, crowned, grinding teeth at the sides and back of the mouth, used for crushing food prior to swallowing. Together with **premolars**, these **teeth** are also known as *cheek teeth*. They are found in **omnivores** and **herbivores** but are replaced by **carnassial** teeth in **carnivores**. See **dental formula**, **dentition**.

molecule The smallest complete part of a chemical **compound** that can take part in a reaction. Within a molecule, **atoms** occur in fixed proportions.

monocotyledons The smaller of the two subsets of flowering plants, the other being **dicotyledons**. The characteristics of monocotyledons are:
(a) one **cotyledon** in the **seed**;
(b) parallel **veins** in **leaves**;
(c) narrow leaves;
(d) scattered **vascular bundles** in **stem**;
(e) **flower** parts usually in threes or in multiples of threes.
Examples are cereals and grasses.

monohybrid inheritance The inheritance of one pair of contrasting characteristics.
For example, in the fruit-fly *Drosophila*, one variety is **pure-breeding** normal-winged, and another is pure-breeding vestigial-winged, the normal-winged **allele** being **dominant** to the vestigial-winged allele.
The approximate ratio in F_2 generation is:

$$\frac{\text{normal}}{\text{vestigial wing}} = \frac{3}{1}$$

When normal-winged is crossed with vestigial-winged, the vestigial trait seems to disappear, only to reappear again to a limited extent in the next generation, suggesting that the **F_1 generation** must have possessed this trait without showing it. A trait such as normal wing which always appears in a cross between contrasting parents is described as **dominant**, while a trait such as vestigial wing which is 'lost' in F_1 generation **progeny**, apparently masked by a dominant trait, is called **recessive**.
For each trait, an organism receives one **gene** from the male **gamete** and one from the female gamete, and the resulting organism i.e. the **zygote**, contains two genes for every trait, but the gametes contain only one (as a result of **meiosis**). If the paired genes for a particular trait are identical, the organism is said to be **homozygous** or *pure* for that trait. When an organism has two different genes for a trait, it is described as **heterozygous** or **hybrid**. Alternative forms of a gene are described as *allelomorphs* or alleles. Hence, if the allele for normal wing is represented by *N* and that for vestigial wing by *n*

NN is homozygous normal wing
Nn is heterozygous normal wing
nn is homozygous vestigial wing

Organisms that are homozygous for a particular trait produce only one type of gamete for that trait, while heterozygous organisms produce two gamete types. The results of **fertilization** can then be worked out using a **Punnett square**. See **backcross**, **incomplete dominance**.

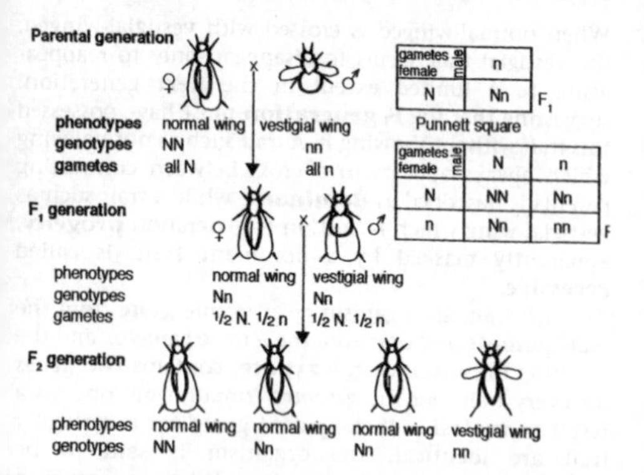

monohybrid inheritance The results of crossing normal-winged female *Drosophila* with vestigial-winged males.

monosaccharides Single **sugar carbohydrates** which are the subunits of more complex carbohydrates, and named on the basis of the number of carbon atoms present. For example:

$C_3H_6O_3$	$C_5H_{10}O_5$	$C_6H_{12}O_6$
triose sugar	pentose sugar	hexose sugar

Hexose sugars are common carbohydrates and include **glucose**.

motile (used of organisms or parts of organisms) Able to move.

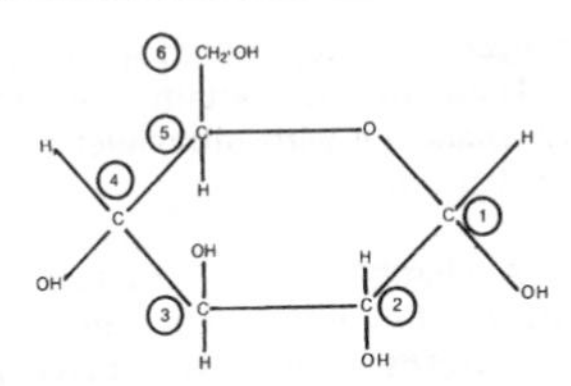

monosaccharides Structure of glucose ($C_6H_{12}O_6$).

mucus A slimy fluid secreted by **goblet cells** in vertebrate **epithelia**. Mucus traps dust and **bacteria** in mammalian air passages, lubricates the surfaces of internal organs, and facilitates the movement of food through the **gut** while preventing the digestive enzymes from reaching and digesting the gut itself.

multicellular (used of an **organism**) Consisting of many **cells**. Most animals and plants are multicellular. Compare **unicellular**.

muscle Animal **tissue** consisting of **cells** which are capable of contraction as a result of **nerve impulses**, thus producing movement, both of the organism as a whole and of internal **organs**. See **antagonistic muscles, involuntary muscles, voluntary muscles**.

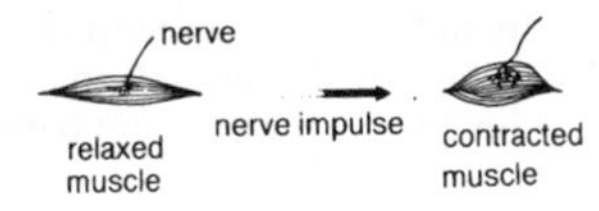

muscle Contraction.

mutagenic agents Factors which speed up the rate of **mutation**. These include certain chemicals such as mustard gas; *irradiation* with ultraviolet rays or X-rays; and atomic radiation.

mutation A change in the structure of **DNA** in **chromosomes**. Mutations occur rarely; when they occur in the **gametes**, or the cells that give rise to them, they are inheritable and most confer disadvantages on the organisms inheriting them. Mutations can result in beneficial **variations** within a **population** which can lead to **evolution**, and although they occur naturally, they can also be induced by exposure to excessive radiation or other **mutagenic agents**.

mutualism A **symbiotic** relationship in which both organisms benefit. For example, the **intestinal bacteria** of **herbivores** digest the **cellulose** of plant **cell** walls, the products of which are then used by the herbivore.

mycorrhiza A **symbiotic** association between **fungi** and the **roots** of certain plants. The fungi provides the plants with **amino acids** and in return receives **carbohydrates**.

nastic movement A growth response by plants to a stimulus that is independent of the direction of the stimulus. An example is the opening and closing of a flower in response to light intensity. See **tropism**.

natural selection The theory proposed by Charles Darwin to explain how **evolution** could have taken

place. Darwin suggested that individuals in a **species** differ in the extent to which they are adapted to their **environment**. Thus, in **competition** for food, etc., the better-adapted organisms will survive, and pass on their favourable **variations**, while the less well adapted will be eliminated.
Over a large number of generations, natural selection can change the characteristics of a **species** and contribute to the development of new species. All the species which exist today are thought to have evolved, by **mutation** and natural selection, from simple organisms which first developed millions of years ago.

nephron A subunit of the vertebrate **kidney**, consisting of a **Bowman's capsule**, **glomerulus**, and renal tubule. See **excretion**, **osmoregulation**.

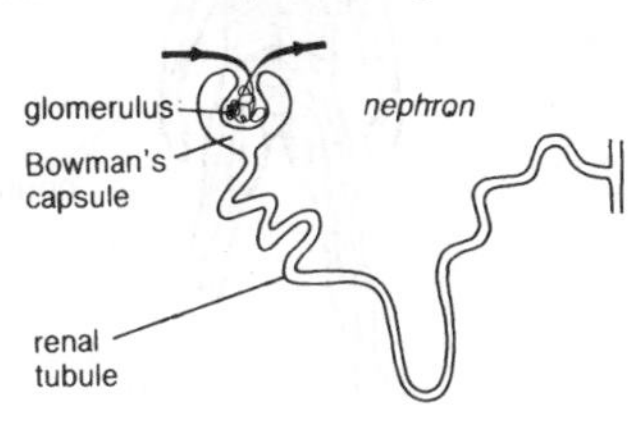

nephron Section through a nephron.

nerve impulses The electrical messages by which information is transmitted rapidly throughout **nervous systems**. Nerve impulses are initiated at **receptor cells** as a result of **stimuli** from the **environment**. In vertebrates, the impulses are conducted to the **central**

nervous system, where they trigger other impulses which are relayed to **effector organs**. See **neurones, synapse.**

nervous system A network of specialized **cells** in **multicellular** animals, which acts as a link between **receptors** and **effectors**, and thus coordinates the animal's activities. In mammals, the nervous system consists of the **brain** and **spinal cord** (which together form the **central nervous system**) and **neurones** connecting all parts of the body.

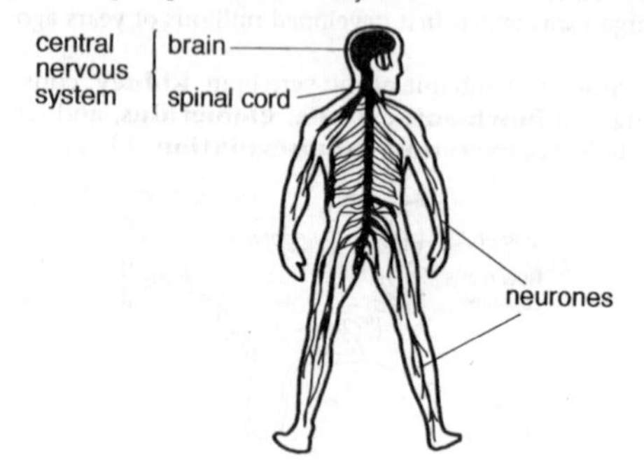

nervous system The main parts of the human nervous system.

neural (used of functions and parts of the body) Related to the **nervous system.**

neurones or **nerve cells** **Cells** which are the basic units of mammalian **nervous systems.**

There are two types of neurone:

(a) *sensory neurones*: these conduct **nerve impulses** from **receptors** to the **central nervous system** (CNS), i.e. from **eyes, ears, skin**, etc.;

(b) *motor neurones*: these conduct **nerve impulses** from the CNS to **effectors**, such as **muscles** and **endocrine glands**.

Each neurone consists of three parts:

(a) a *cell body* containing **cytoplasm** and **nucleus** and forming the *grey matter* in the **brain** and **spinal cord**;

(b) fibres which carry nerve impulses into cell bodies. In sensory neurones, this fibre is a single *dendron* while in motor neurones there are numerous *dendrites*. In the CNS such fibres form *white matter*;

(c) fibres called **axons** which carry nerve impulses from cell bodies.

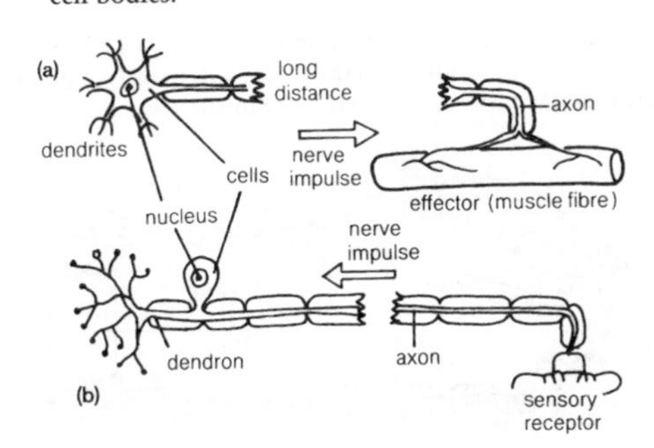

neurones The structure of (a) motor neurone and (b) sensory neurone.

neutralize See **alkali**.

neutron See **atom**.

niche The status or way of life of an organism within a **community**. For example, a **herbivore** and a **carnivore** may share the same **habitat** but their different feeding methods mean that they occupy different niches.

nitrification The conversion of **organic** nitrogen compounds by nitrifying **bacteria** in the **soil**, for example, ammonia into nitrates which can be absorbed by plants. Ammonia is first converted to nitrites by *Nitrosomonas* **bacteria** and the nitrites to nitrates by *Nitrobacter* **species**. See **nitrogen cycle**.

nitrogen cycle The circulation of the element nitrogen and its compounds in nature, caused mainly by the **metabolic** processes of living organisms.

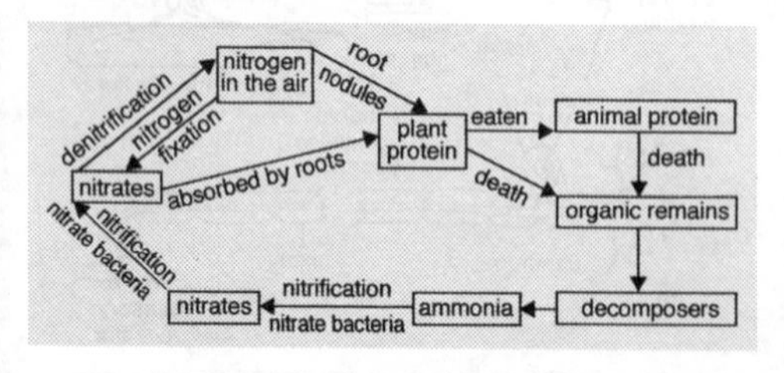

nitrogen cycle The main steps.

nitrogen fixation The conversion of atmospheric nitrogen by certain **microorganisms** into **organic** nitrogen compounds. Nitrogen-fixing **bacteria** live either in **soil**, air, or within the **root nodules** of *leguminous plants* (peas, beans, clover). The activity of these organisms, such as *Azotobacter* and *Rhizobium*, enriches the soil with nitrogen compounds. See **nitrogen cycle**, **root nodule**.

normal distribution curve A bell-shaped curve obtained when *continuous variation* is measured in a **population**. See **variation**.

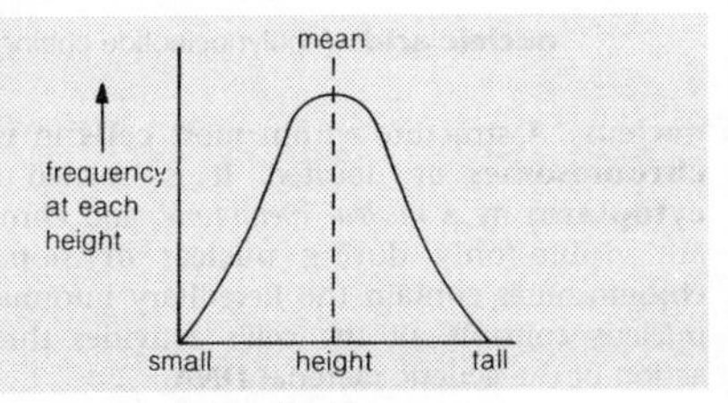

normal distribution curve

nuclear division See **meiosis**, **mitosis**.

nucleic acids **Organic compounds** found in all

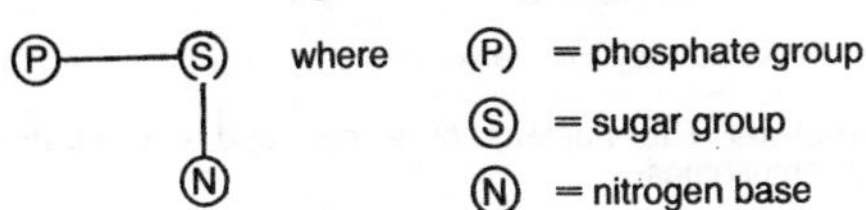

nucleic acids Structure of a single nucleotide.

living organisms, particularly associated with the **nucleus** of the **cell**, and consisting of subunits called *nucleotides*.
The sugar group of one nucleotide can combine with the phosphate group of another to form a *polynucleotide* chain. Such polynucleotide chains are the basis of nucleic acid structure. See **DNA**, **RNA**.

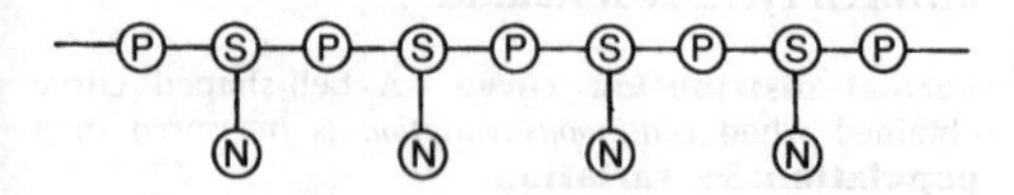

nucleic acids A polynucleotide chain.

nucleus A structure within most **cells** in which the **chromosomes** are located. It is isolated from the **cytoplasm** by a *nuclear membrane*, and chromosomes are visible only during nuclear division. As the chromosomes contain the hereditary information, the nucleus controls all the cell's activities through the action of the genetic material **DNA**.

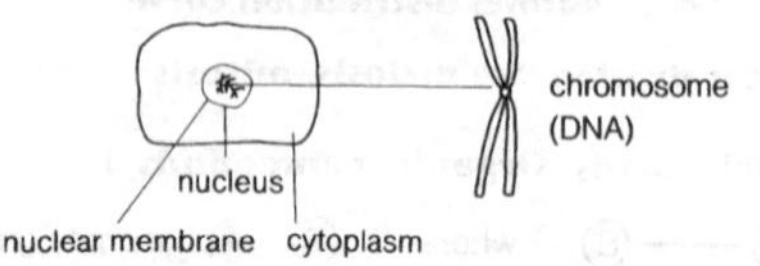

nucleus The nucleus of a cell and one of its chromosomes.

nutrient agar plate See **agar**.

nymph The juvenile form of certain insects which resembles the **imago** except that it is smaller, wingless and sexually immature. See **metamorphosis**.

oesophagus A region of the **alimentary canal**, connecting the mouth with the digestive areas. In vertebrates, it runs between the **pharynx** and the **stomach** and transports food by **peristalsis**. See **digestion**.

oestrogen A **hormone** secreted by vertebrate **ovaries** which stimulates the development of **secondary sexual characteristics** in female mammals and is important in the **menstrual cycle**.

olfactory (used of parts of the body and functions) Related to the sense of **smell**.

omnivore An animal which feeds on both plants and animals. Omnivores include humans, whose **dentition**, like other omnivores, contains both **herbivore** and **carnivore** features, consisting of biting, ripping and grinding teeth, which suit the mixed diet. See **dental formula, teeth**.

optic (used of parts of the body) Related to the **eye** and its functions.

optic nerve **A cranial nerve** of vertebrates conducting **nerve impulses** from the **retina** of the **eye** to the **brain**.

oral (used of parts of the body and functions) Related to the mouth.

oral hygiene The maintenance of healthy conditions in the human mouth in order to reduce tooth decay. **Bacteria** in the mouth convert food particles stuck to **teeth** into **acid** which attacks the teeth and causes decay. Regular brushing with toothpaste reduces the likelihood of tooth decay.

order A unit used in the **classification** of living organisms, consisting of one or more **families**.

organ A collection of different **tissues** in a plant or animal which form a structural and functional unit. Examples are the **liver** and a plant **leaf**.
Different organs may then be associated together to constitute a *system*, e.g. the digestive system.

cells→tissues→organs→systems

organelle A structure found in the **cytoplasm** of **cells**. Examples are **mitochondria**, **chloroplasts**. See **cell differentiation**.

organ of Corti See **cochlea**.

organic compounds Compounds containing the element carbon, found in all living organisms. The major organic compounds are **carbohydrates**, **fats**, **nucleic acids**, **proteins** and **vitamins**. See **inorganic compounds**.

osmoregulation Control of the osmotic pressure, and therefore the **water** content, of an organism. See **osmosis**.

In terrestrial organisms, water is gained from food and drink and as a by-product of **respiration**, and is lost by sweating, in exhaled air, and as urine.
Water and mineral salt balance in terrestrial animals is mainly under the control of the **kidneys**. For osmoregulation in fish, see diagram.

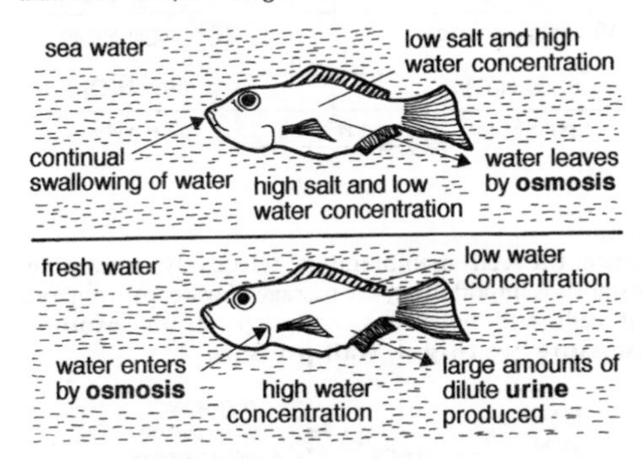

osmoregulation Water balance in a sea fish (above) and a freshwater fish (below).

osmosis The **diffusion** of **solvent** (usually **water**) particles through a **selectively permeable membrane** from a region of high solvent concentration to a region of lower solvent concentration. See **solution**.
Examples of selectively permeable membranes are (a) the **cell membrane**, and (b) *visking tubing* (used in *dialysis* – see **kidney machine**). Such membranes are thought to have tiny pores which allow the rapid passage of small water particles, but restrict the passage of larger *solute*

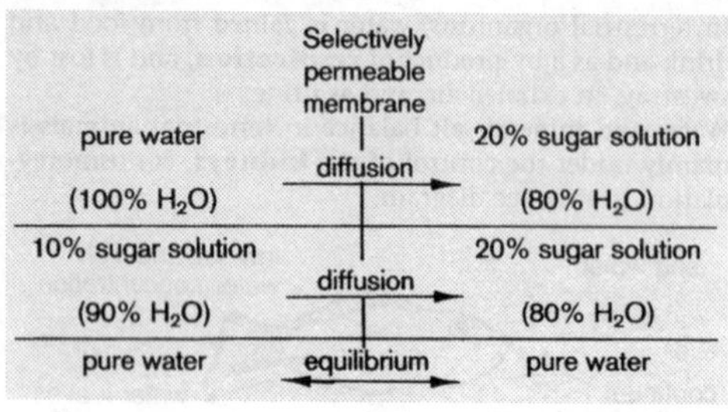

osmosis

particles.
Since the cell membrane is selectively permeable, osmosis is important in the passage of water into and out of cells and organisms, and depends on **osmotic pressure**. See **turgor, wall pressure**.

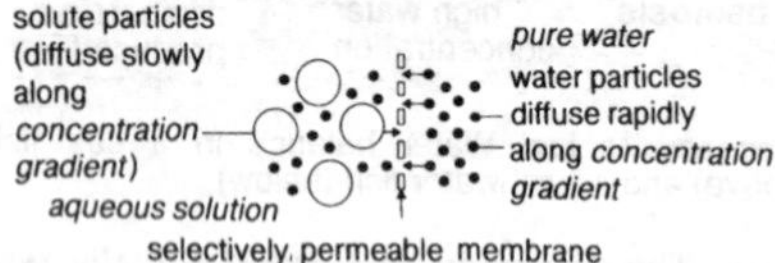

osmosis Passage of solvent (water) particles through a selectively permeable membrane.

osmotic pressure The pressure exerted by the osmotic movement of **water** which can be demonstrated in an *osmometer*.
Water moves by **osmosis** into the sugar solution via the visking tubing, causing the liquid level in the tube to

rise. Osmotic pressure depends on the relative **solute** concentrations of the **solutions** involved. The osmotic pressure that a solution is capable of developing is called its *osmotic potential*, but is only realized in an osmometer. See **turgor**, **wall pressure**.

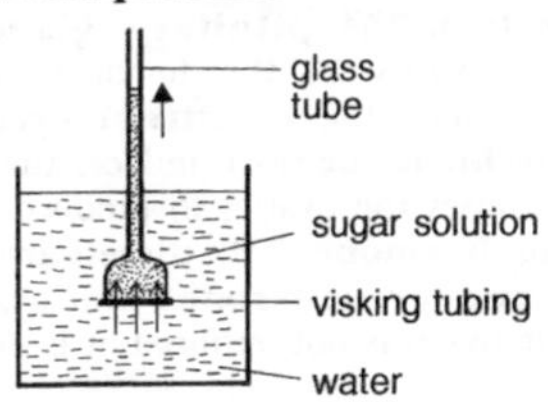

osmotic pressure An osmometer.

ossicles The three tiny linked **bones** in the mammalian middle ear. See **ear**.

oval window A membrane separating the middle ear and inner ear in mammals. See **ear**.

ovary **1.** A hollow region in the **carpel** of a **flower**, containing one or more **ovules**. See **fertilization**.
2. The reproductive **organ** of female animals. In vertebrates, there are two ovaries which produce the **ova** and also release certain sex **hormones**. See **fertilization**, **ovulation**.

oviduct A tube in animals which carries **ova** from the **ovaries**. In mammals there are two oviducts leading to the **uterus**, and **fertilization** occurs within the oviduct.

ovulation The release of an **ovum** from a mature **Graafian follicle** on the surface of a vertebrate **ovary**, from where it passes into the **oviduct** and then into the **uterus**.

Ovulation in the human female: ovulation is controlled by **hormones** from the **pituitary gland** and the sequence of events in the female's reproductive behaviour is called the **menstrual cycle**. **Follicle stimulating hormone** (FSH) induces the maturation of ova and causes the ovaries to produce **oestrogen**. **Luteinizing hormone** (LH) triggers ovulation and also the release of **progesterone** by the ovaries.

If the mature ovum is not fertilized, it is expelled with

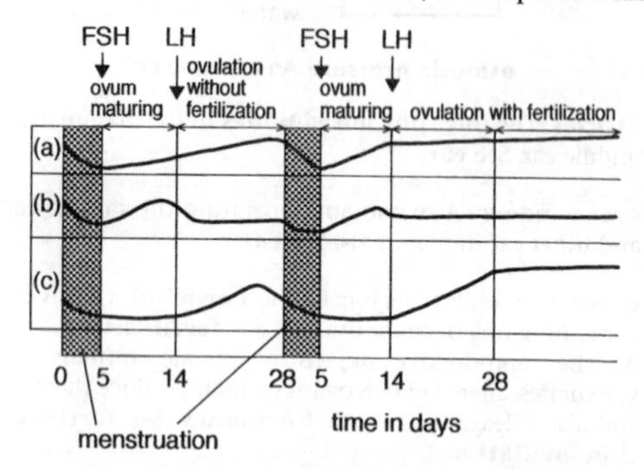

ovulation The main features of the human menstrual cycle: (a) variation in thickness of wall of uterus; (b) oestrogen level in blood (stimulates repair of the uterus wall); (c) progesterone level in blood (prepares uterus for implantation).

the new uterus lining and some **blood** via the **vagina**, a process called **menstruation**.

ovule A structure in flowering plants which develops into a **seed** after **fertilization**. See **carpel**.

ovum (*pl.* **ova**) An unfertilized female **gamete** produced in the **ovary** of many animals, and containing a **haploid nucleus**. See **fertilization**, **meiosis**, **ovulation**.

oxygenated blood See **artery**.

oxygen debt A deficit of oxygen which occurs in **aerobes** when work is done with inadequate oxygen supply. For example, in mammalian **muscle** during **exercise**, the oxygen supply may be insufficient to meet the **energy** demand. When this happens, the **cells** produce energy by **anaerobic respiration**, **lactic acid** being a by-product.
The accumulation of lactic acid causes muscle fatigue but is eventually reduced as oxygen intake returns to normal after the period of exercise. This shortfall of

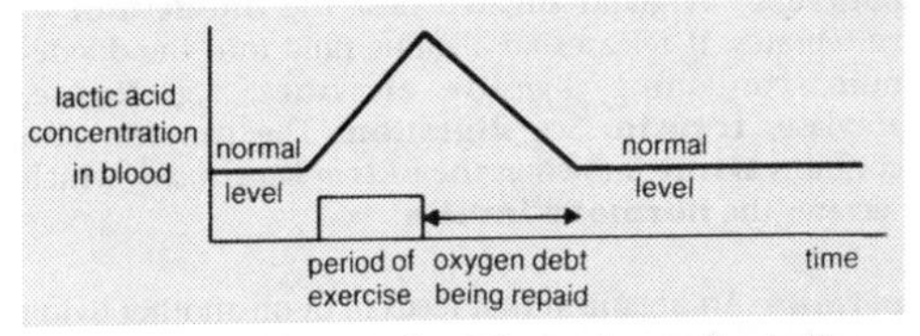

oxygen debt The effect of exercise on the lactic acid concentration of the blood.

oxygen must be repaid by increased oxygen intake (panting). See **blood, respiration**.

ozone layer A layer of the atmospheric gas ozone (O_3) which surrounds planet Earth. It is thought that the layer is being depleted as the result of reacting with *chlorofluocarbons* (CFC) which are gases released by aerosol cans. This may allow more ultraviolet (UV) radiation from the Sun to reach the Earth's surface, causing skin cancer and disturbing weather patterns.

pacemaker **1.** A structure in the right **atrium** of the **heart**, which initiates the **heartbeat**.
2. An electronic device implanted in the heart to stimulate and regulate the heartbeat.

palaeontology See **fossil**.

palisade mesophyll The main **tissue** carrying out **photosynthesis** in the **leaf**, situated below the upper **epidermis**, and containing many **chloroplasts**.

pancreas A gland situated near the **duodenum** of vertebrates. It releases an alkaline fluid into the duodenum, containing digestive **enzymes**, e.g. **lipase**, amylase, **trypsin**. See **digestion**. The pancreas also contains **tissue** known as the *islets of Langerhans*, which secretes the **hormone insulin**.

parasite An organism that feeds in or on another living organism which is called the *host*, and which does not benefit and may be harmed by the relationship.

Parasites of man include fleas, lice and tapeworms. See **ectoparasite, endoparasite.**

parasitism See **symbiosis.**

parental generation The first organisms crossed in a breeding experiment, producing progeny known as the **F_1 generation**. See **monohybrid inheritance.**

patella A **bone** over the front of the knee **joint** in many vertebrates. In humans it is the kneecap. See **endoskeleton.**

pathogen A term used to describe an organism causing disease in another **species**. Examples are **viruses, bacteria** and tapeworms. See **parasite.**

pathogenic microorganisms **Microorganisms** which cause disease. Such organisms enter the body by various routes: contaminated food, inhaled air, insect bites, **skin** wounds. However, the body has mechanisms for preventing infection:
(a) the **acid pH** of the **stomach** kills many organisms;
(b) the formation of **blood clots** at wounds restricts entry;
(c) **white blood cells** destroy microorganisms by **phagocytosis**;
(d) **antibodies** in the **blood** neutralize the poisons produced by microorganisms;
(e) **cilia** and **mucus** clear the air passages.
If these natural defence mechanisms are overcome, the infection can then be treated using **antibiotics.** Prevention is also possible by **immunization.**

pectoral Relating to that part of the body at the **anterior** end of the trunk of an animal (e.g. the shoulders) to which the forelimbs are attached.

pelvic Relating to that part of the body of an animal which forms the lower abdomen and to which the hindlimbs are attached.

penicillin The first **antibiotic**, discovered by Alexander Fleming in 1928.

penis An **organ** in mammals by which the male **gametes** (**spermatozoa**) are introduced into the female body. It also contains the **urethra** through which **urine** is discharged. See **fertilization**.

pentadactyl limb A limb with five digits, characteristic of **tetrapod** animals. There is a basic arrangement of bones that is modified in many species. See **endoskeleton**.

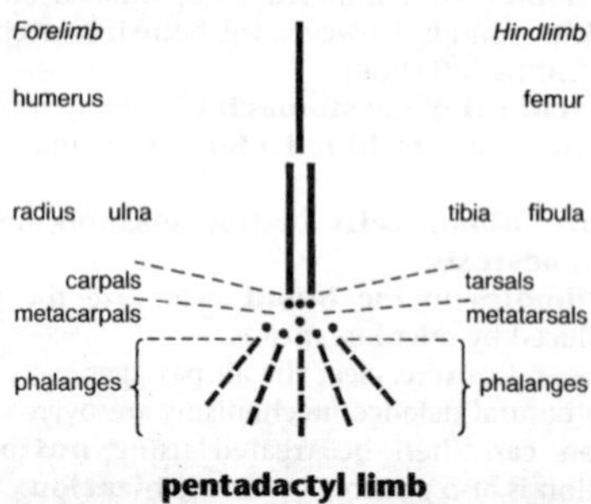

pentadactyl limb

pepsin A **protease enzyme** secreted by the wall of the vertebrate **stomach**, along with hydrochloric **acid**. The acid provides a suitable **pH** for pepsin which digests long protein chains into shorter chains of **amino acids** called **peptides**.

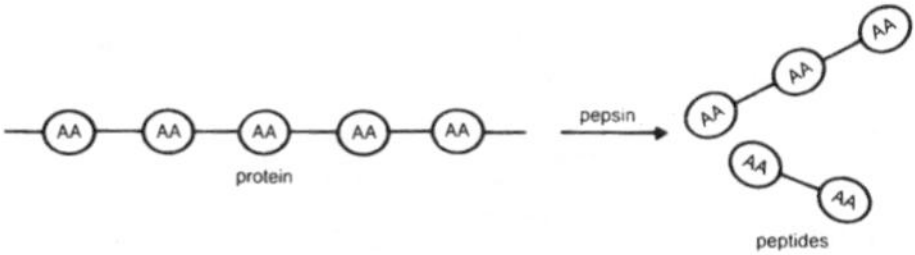

pepsin The action of pepsin on protein.

peptide A compound consisting of two or more **amino acids** linked between the amino group of one and the **acid** group of the next. The link between adjacent amino acids is called a *peptide bond*, and when many amino acids are joined in this way, the whole complex is called a *polypeptide*, which is the basis of **protein** structure.

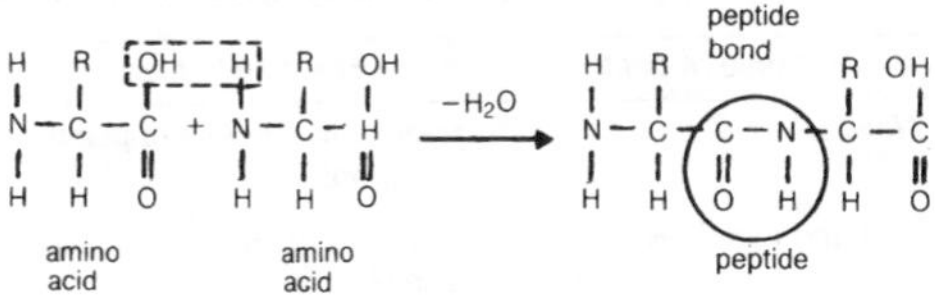

peptide Two amino acids are joined in a peptide bond.

peristalsis Waves of muscular contraction that pass along tubular organs and cause movement of their contents. For example, in mammals, peristalsis occurs in the **alimentary canal** and also in the **ureters** and **oviducts**. Peristalsis is caused by the rhythmic and coordinated contraction and relaxation of both circular and longitudinal **involuntary muscles**. See **antagonistic muscles**.

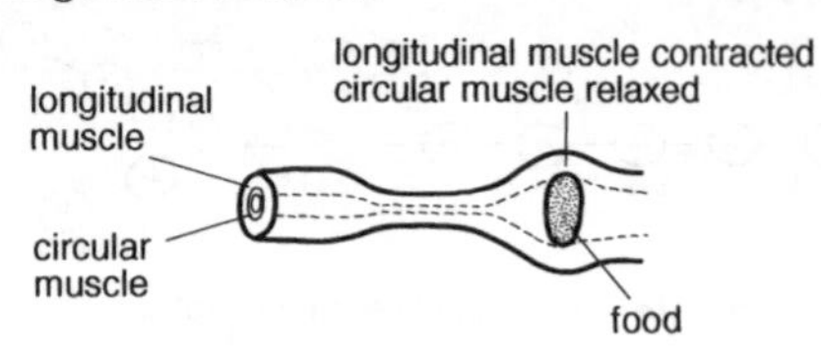

peristalsis

pest Any living organism which is considered to have a detrimental effect on humans.
Methods used to combat pests include:
(a) spraying with chemicals (**pesticides**);
(b) using natural **predators** against the pest;
(c) introducing **parasites** and **pathogens** to the pest

Example of pest	Effect on humans
Weeds, locusts	Reduce the growth of plants and crops.
Foot and mouth virus	Causes disease in domestic animals.
Woodworm, wet rot fungus	Damages buildings.
Mosquitoes, lice	Transmit human disease.

pest Some pests and the damage they cause.

population;

(d) introducing **sterile** individuals to the pest population, thus reducing reproductive capacity.

pesticide A chemical compound, often delivered in a spray, which kills pests or inhibits their growth.
Examples:

(a) *herbicides*: weedkillers such as paraquat;
(b) *fungicides*: seed dressings such as organo-mercury compounds;
(c) *insecticides*: fly sprays used in the home; sprays released from aircraft against locusts. DDT (now banned in Britain) has been used against mosquitoes and lice.

The disadvantages of pesticides are:

(a) they may kill organisms other than the target pest;
(b) the concentration of a pesticide increases as it passes through a **food chain**;
(c) some decompose slowly and may accumulate into harmful doses within other organisms;
(d) by killing off susceptible organisms, they allow resistant individuals to grow and multiply with reduced **competition**.

pH A measure of the degree of *acidity* or *alkalinity* of a **solution**.

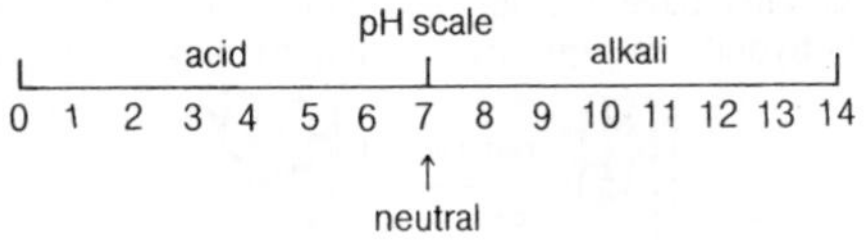

pH scale
acid | alkali
0 1 2 3 4 5 6 7 8 9 10 11 12 13 14
↑
neutral

pH The pH scale.

phagocytosis The process by which certain **cells** (*phagocytes*) surround and engulf a food particle which is then digested. Phagocytosis is the feeding method employed by some **unicellular** protozoans, e.g. *Amoeba*. It is also one of the methods by which **white blood cells** destroy invading **microorganisms**.

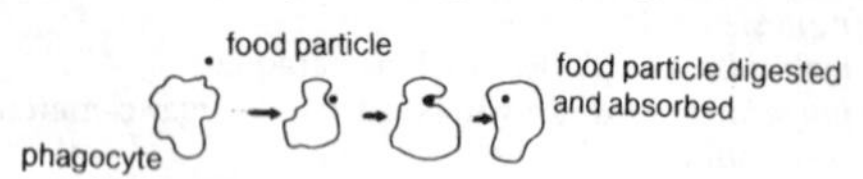

phagocytosis The phagocyte engulfs and digests a food particle.

pharynx A region of the vertebrate **alimentary canal** between the **mouth** and the **oesophagus**. In humans it is the back of the nose and throat, and when it is stimulated by food, swallowing is initiated.

phenotype The physical characteristics of an organism resulting from the influence of **genotype** and **environment**. See **monohybrid inheritance**.

phloem **Tissue** within plants which transports **carbohydrate** from the **leaves** throughout the plant. Phloem consists of tubes which are formed from columns of living **cells** in which the horizontal cross-walls have become perforated. This allows the carbohydrate in aqueous solution to move from one

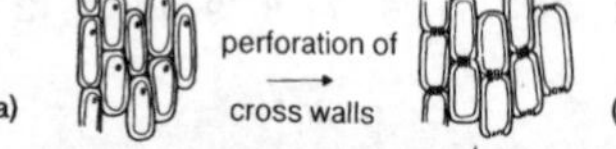

phloem Phloem cells (a) become sieve tubes (b).

phloem cell into the next and thus through the plant. Because of their structure, phloem tubes are also called *sieve tubes*. See **companion cell**, **leaf**, **root**, **stem**, **translocation**.

photoperiod The length of daylight which is important in many plant and animal **responses**, such as flowering in plants; migration in birds. See **rhythmical behaviour**.

photoreceptor A **sense organ** which is stimulated by light, such as the **eye**.

photosynthesis The process by which green plants make **carbohydrate** from carbon dioxide and **water**. The **energy** for the reaction comes from sunlight which is absorbed by the **chlorophyll** within **chloroplasts**, and oxygen is evolved as a by-product. Overall reaction:

carbon + water dioxide	light energy → chlorophyll	carbohydrate + oxygen
$6CO_2 + 6H_2O$		$C_6H_{12}O_6 + 6O_2$

Photosynthesis is in fact a two-stage reaction involving:

(a) the *light reaction* in which light energy is used to split water into hydrogen (which passes to the next stage) and oxygen (which is released);

(b) the *dark reaction* in which the hydrogen from the light reaction combines with carbon dioxide to form carbohydrates.

It is the source of all food and the basis of **food chains**, while the release of oxygen replenishes the oxygen content of the atmosphere.

phototropism **Tropism** relative to light. Plant shoots are *positively phototropic*, i.e. they grow towards light.

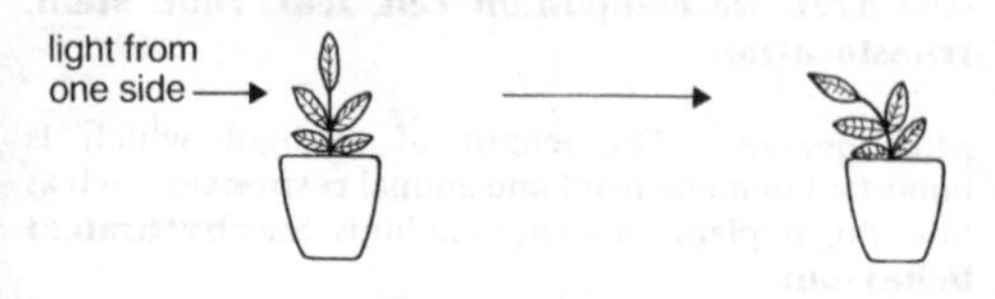

phototropism

phylum A unit used in the **classification** of living organisms, consisting of one or more **classes**. The term **division** is often substituted in plant classification.

phytoplankton See **plankton**.

pinna A flap of **skin** and **cartilage** at the outside end of the mammalian outer ear. See **ear**.

pitfall trap A trap used to collect organisms living on or just below the **soil** surface and from leaf litter. Examples of organisms caught are beetles and centipedes. See diagram.

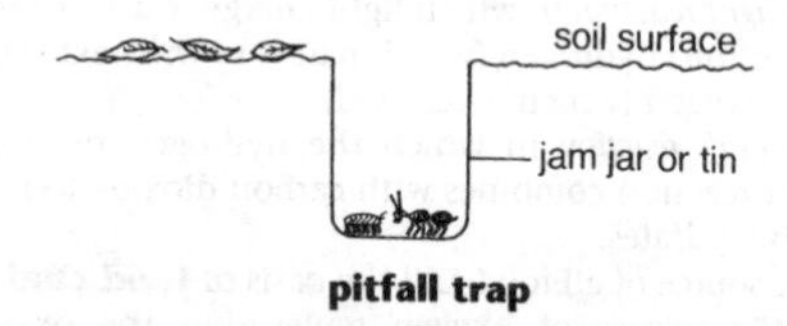

pitfall trap

pituitary gland An **endocrine gland** at the base of

the vertebrate **brain**. It produces numerous **hormones** including **antidiuretic hormone** and **follicle-stimulating hormone**, many of which regulate the activity of other endocrine glands. The pituitary gland's own secretion is in many cases regulated by the brain.

placenta An **organ** that develops during **pregnancy** in the mammalian **uterus** and forms a close association between maternal and foetal **blood** circulations. The placenta allows passage of food and oxygen to the **foetus** and removes carbon dioxide and **urea**.

plankton Microscopic animals (**zooplankton**) and plants (**phytoplankton**) which float in the surface waters of lakes and seas. Plankton are important as the basis of aquatic **food chains**.

plant growth substances Chemical compounds such as **auxins**, involved in many plant processes including **tropisms**, **germination** etc. The term 'plant **hormone**' was formerly used.

plasma The clear fluid of vertebrate **blood** in which the blood **cells** are suspended. It is an aqueous **solution** in which are dissolved many compounds in transit around the body. Examples:

Compound	
carbon dioxide urea	waste products
glucose amino acids	digested foods
hormones	
plasma proteins	
sodium chloride	

plasma proteins **Proteins** dissolved in the **plasma** of vertebrate **blood**. Examples are **antibodies**, **fibrinogen**, and some **hormones**.

plasmid A circular group of **genes** found in **bacteria cells**, independent of the bacterial **chromosome**. See **genetic engineering**.

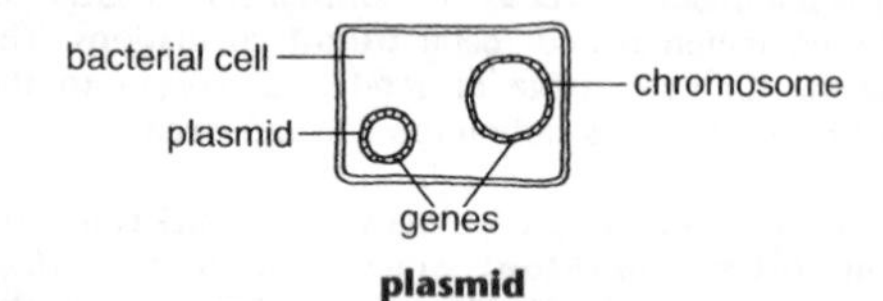

plasmid

plasmolysis The loss of **water** from a plant **cell** by **osmosis** when the cell is surrounded by a **solution** whose water concentration is less than that of the cell **vacuole** (for example, a strong **sugar** or salt solution). Osmosis causes water to pass out of the cell, making the vacuole shrink, resulting in the pulling away of the **cytoplasm** from the cell wall.
Plasmolysis can be induced and reversed experimentally, but continued plasmolysis results in cell death. This osmotic death rarely occurs naturally but can result from

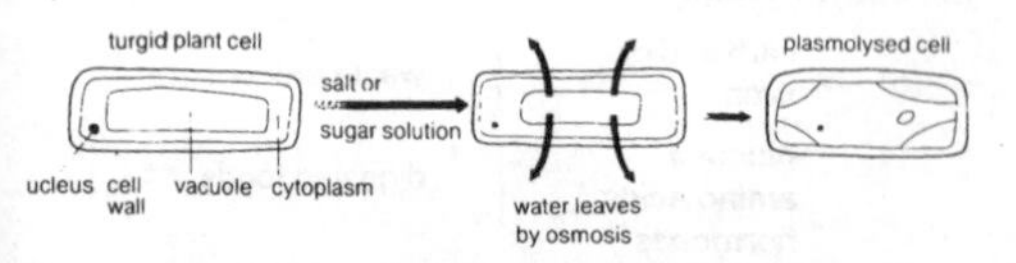

plasmolysis

adding excessive **fertilizer** to plants which induces plasmolysis, or what is called 'plant burning'.

platelet The smallest **cell** of mammalian **blood**, involved in **blood clotting**.

pleural membranes The two **membranes** that cover the outside of the **lungs** and line the inside of the **thorax** in mammals, and that secrete pleural fluid between them, so facilitating **breathing** movements.

plumule The leafy part of the embryonic **shoot** of **seed** plants. See **germination**.

pollen Reproductive **cells** of flowering plants, each containing a male **gamete**. Pollen grains are adapted to their mode of transfer, which may be either by insects or by wind.

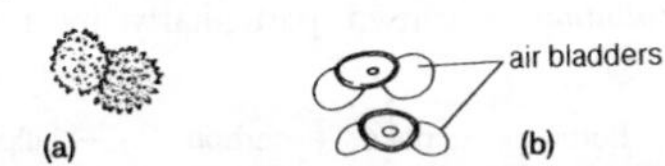

pollen (a) Spiky and sticky for insect pollination. (b) Smooth and light for wind pollination.

pollination The transfer of **pollen** grains from **stamens** to **carpels** in flowering plants. Pollination within the same **flower** or between flowers on the same plant is called *self-pollination*. Pollination between two separate plants is called *cross-pollination*. Normally male and female parts of the same plant do not mature simultaneously, favouring cross-pollination with a consequent mixing of **chromosomes**, which can lead to **variation**. Pollen is transferred on the bodies of insects or by the wind. Flowers are adapted to favour one particular method of transfer:

(a) *insect pollination*: insects visit flowers to drink or collect nectar, attracted by the colour or scent of the flower. Their bodies become dusted with pollen, some of which may adhere to the **stigmas** of subsequent flowers which they visit;

(b) *wind pollination*: pollen grains carried by the wind must be produced in much higher numbers than those carried by insects, to compensate for loss during transfer. See **fertilization**.

pollution The addition of any substance to the **environment** which upsets the natural balance. Pollution has resulted mainly from industrialization which is largely based on **fossil fuel** burning and which caused migration from the land to towns and cities.

(a) *Air pollution* is caused particularly by fossil fuel burning.

coal	burning ⟶	smoke	+ carbon dioxide	+ sulphur dioxide
petrol	burning ⟶	smoke	+ carbon monoxide	+ oxides of nitrogen + lead

Air pollutants such as smoke and sulphur dioxide cause irritation in the human respiratory system and may accelerate diseases such as bronchitis and lung cancer.

(b) *Water pollution* results from the intentional or accidental addition of materials to either fresh

water or sea water. The pollutants originate from industrial and agricultural practices and also from the home. Examples are mine and quarry washings, **acids**, **pesticides**, oil, radioactive wastes, fertilizers, detergents, sewage, hot water (from power stations). Some pollutants, such as pesticides, may poison aquatic organisms, while organic pollutants, such as sewage, cause an increase in the **microorganism** population in the water with a resulting decrease in dissolved oxygen levels, making the water unfit for many organisms.

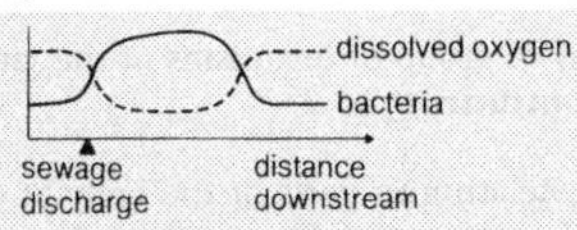

pollution The effect of sewage discharge on oxygen levels in water.

polyploidy A **mutation** in which an organism has extra sets of **chromosomes**. Polyploidy is more common in plants than in animals and often confers advantages such as increased disease resistance.

polysaccharides **Carbohydrates** consisting of long chains of **monosaccharides** linked together by bonds. **Glucose** units can be linked in different ways to form several polysaccharides, such as **starch**, **glycogen** and **cellulose**.

polysaccharides A polysaccharide chain.

pooter A device used to collect small insects by using suction.

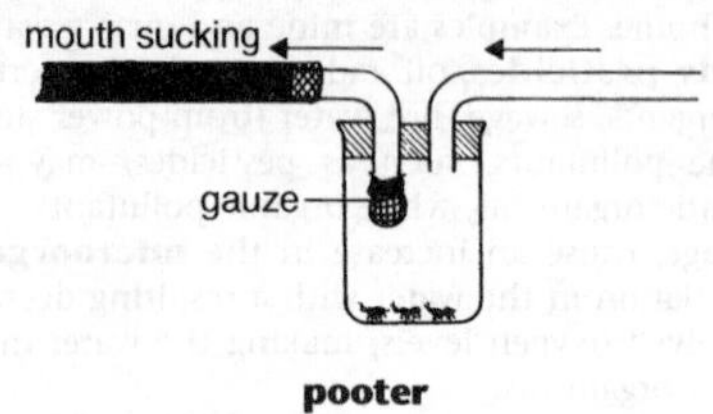

pooter

population A group of organisms of the same **species** within a **community**.

posterior Relating to parts of the body at or near the hind end of an animal. Compare **anterior**.

predator An animal that feeds on other animals which are called the **prey**, and which it catches; i.e. a predator is a **food consumer** but is not a **parasite**. The relationship between predator and prey can have dramatic effects on their numbers.

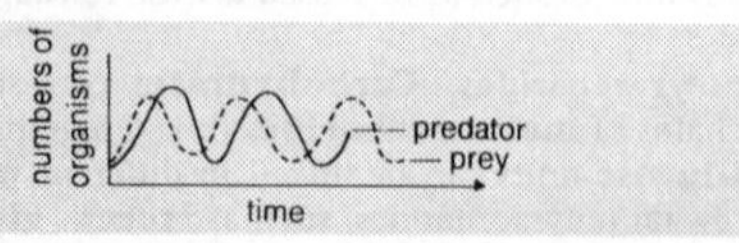

predator A typical predator–prey relationship.

pregnancy or **gestation period** The time from **conception** to **birth** in mammals. Human pregnancy lasts about 40 weeks, the **embryo** developing in the

uterus after **implantation**. Finger-like structures (**villi**) grow from the embryo and develop into the **placenta**.

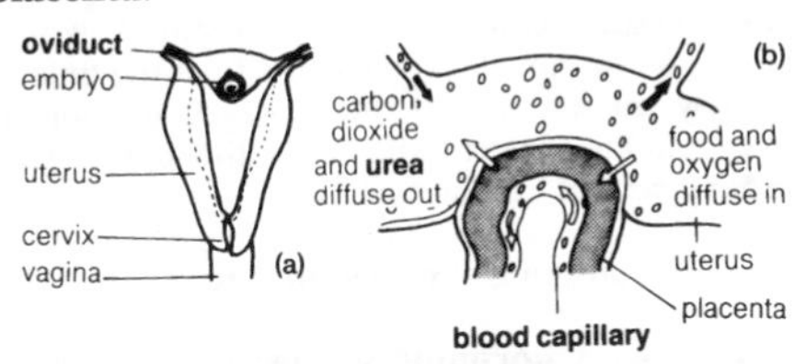

pregnancy (a) Implanted embryo. (b) Relationship between uterus and placenta.

During pregnancy the cells of the embryo continually divide and differentiate, and the growing embryo (**foetus**) becomes suspended in a water sac, the **amnion**. The placenta extends into the **umbilical cord** which connects with the **abdomen** of the foetus.

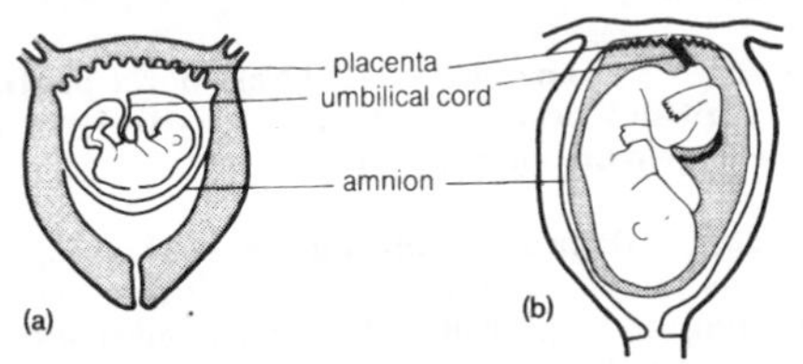

pregnancy (a) Human foetus after twenty weeks. (b) Human foetus just before birth.

premolars **Teeth** found in front of **molars**.

prey An animal hunted for food by another animal. See **predator**.

primary sexual characteristics Features which distinguish between males and females from the time of birth. They do not include those which develop at **puberty** and are characteristic of adulthood. Compare **secondary sexual characteristics**.

progeny The offspring of **reproduction**.

progesterone A **hormone** secreted by mammalian **ovaries**, which prepares the **uterus** for **implantation** and prevents further **ovulation** during **pregnancy**.

propagation See **vegetative reproduction**.

proprioreceptor A **receptor** which is stimulated by change in position of the body, e.g. stretch receptors in **muscle** fibres.

protease Any **enzyme** which breaks down **protein** into **peptides or amino acids**, by **hydrolysis**, for example, **pepsin** and **trypsin**.

proteins **Organic compounds** containing the elements carbon, hydrogen, oxygen and nitrogen and consisting of long chains of subunits called **amino acids**. These chains may then be combined with others, and folded in several different ways, with various types of chemical bonding between chains and parts of chains, giving very large and complex molecules.
Proteins are the 'building blocks' of **cells** and **tissues**,

being important constituents of **muscle**, **skin**, **bone** etc. Proteins also play a vital role as enzymes while some **hormones** are protein in structure. See **peptide**.

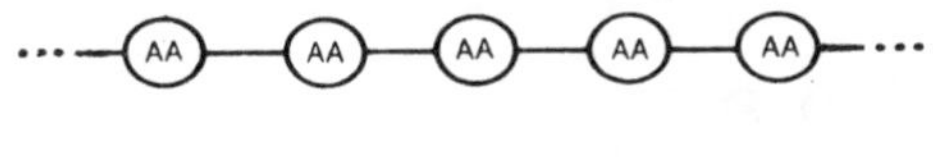

proteins Amino acid chains combine to form large and complex molecules.

protein synthesis The synthesis of protein molecules from their constituent **amino acids**. The information for the construction of a protein with the correct sequence of amino acids is carried in the arrangement of nitrogen base pairs in **DNA**. This **genetic code** is transcribed exactly into **RNA**. (See diagrams.) The first stage is the unzipping of a DNA molecule.
Assume that the N-base sequence of the upper strand is to be transcribed. Nucleotide raw materials (see **nucleic acids**) bond to the appropriate N-base along the DNA strand.
The finished RNA molecule separates from the DNA which has acted as a template. The specific DNA code is now imprinted on the RNA molecule in the form of the corresponding bases. This code represents the information required for protein synthesis. The genetic code 'names' each amino acid by a sequence of 3 adjacent N-bases in DNA and then RNA. These *triplet codes* have been identified for the 20 to 24 amino acids found in living organisms.
The RNA containing the coded triplets is called **messen-**

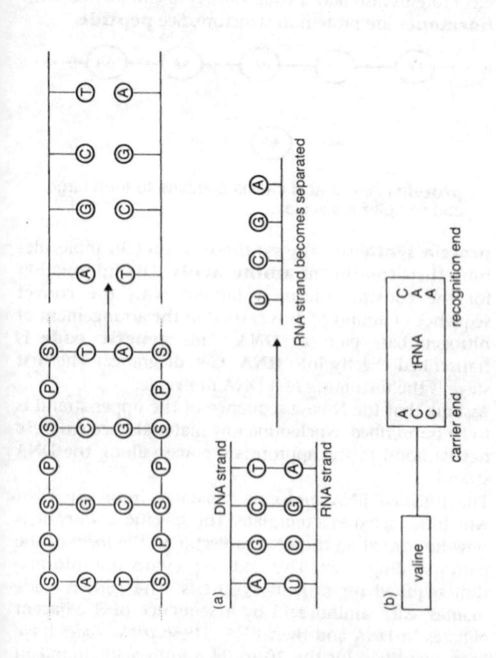

(a)
S P S P S P S
A G C T
T C G A
S P S P S P S
DNA strand
A G C T
U C G A
RNA strand
U C G A
RNA strand becomes separated
(b)
valine
A C C
tRNA
C A A
carrier end
recognition end

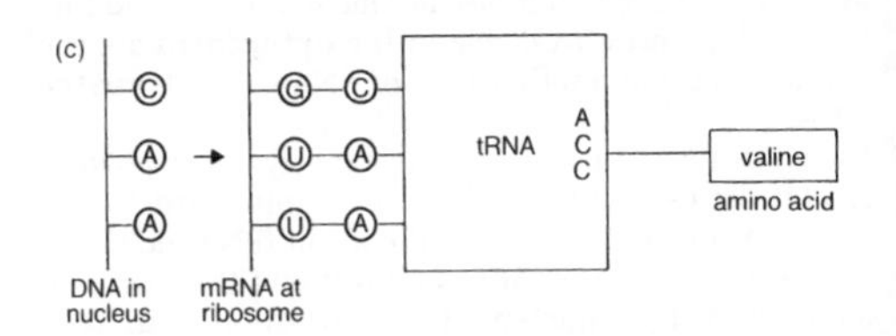

protein synthesis (a) RNA separates from DNA. (b) A tRNA molecule. (c) tRNA bonds with mRNA at a ribosome.

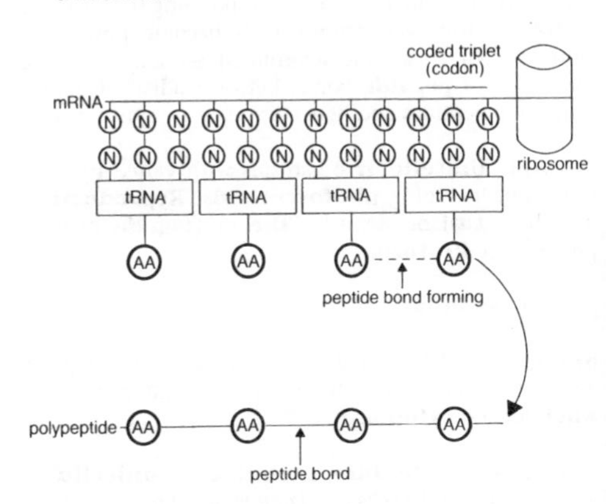

protein synthesis Amino acids bond at a ribosome to form polypeptide.

ger RNA. The mRNA molecules move towards and line up at **ribosomes**. Available in the **cytoplasm** are the amino acids and another type of RNA called **transfer RNA**.

Each tRNA molecule is a comparatively short polynucleotide at each end of which is an important N-base triplet. At one end, the 'carrier' end, all tRNA molecules have the same triplet: ACC, and this end links with an amino acid. The other end, the 'recognition' end, has a triplet which is specific for a particular amino acid, e.g. the amino acid valine has the specific triplet CAA.

When the tRNA arrives at a ribosome, the specific triplet will be able to bond only to a corresponding triplet along mRNA. In this way, amino acids become positioned along mRNA in a code-determined sequence. By the formation of **peptide** bonds between adjacent amino acids, polypeptides and hence proteins are synthesized.

Protista **Unicellular** organisms with varied types of nutrition, including **photosynthesis**. **Reproduction** may be sexual or asexual. **Cilia** or **flagella** may be present. See **Protozoa**.

proton See **atom**.

protoplasm All the material within and including the cell membrane (see **cell**), i.e. protoplasm consists of **nucleus** and **cytoplasm**.

protozoa A **phylum** consisting of **unicellular** animals such as *Amoeba* and *Paramecium*. Protozoans live in a wide variety of **habitats** including stagnant water and **faeces**. They feed on **bacteria** and some cause disease in humans, such as dysentery and malaria.

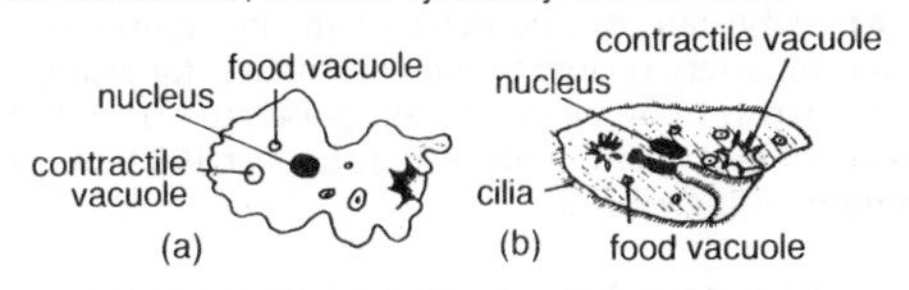

protozoa (a) *Amoeba* (b) *Paramecium.*

puberty The period in the human **life cycle** when a person becomes sexually mature. In females, usually at the age of 10–14 years, **ovulation** begins, and in males at about the same age production of **spermatozoa** starts. In both sexes, during puberty the **secondary sexual characteristics** develop.

pulmonary Related to the **lungs** and **breathing**. The term is applied to **organs**, **tissues** and parts of the body. See **pulmonary vessels**, **heart**, **circulatory system**.

pulmonary vein See **pulmonary vessels**, **heart**, **circulatory system**.

pulmonary vessels **Blood vessels** in mammals, which because of their special functions, do not obey the general rule that **arteries** carry oxygenated blood, and **veins** carry deoxygenated blood. The *pulmonary artery* carries deoxygenated blood from the right **ventricle** to the **lungs**, and the *pulmonary vein* carries oxygenated blood from the lungs to the left **atrium**. See **heart**, **circulatory system**.

pulse rate The regular beating in **arteries** due to

rhythmic movement of **blood** resulting from **heart-beat**. Pulse rate can be detected in the human body where an artery is close to the skin surface, for example, at the wrist. In an adult human, pulse rate varies from about 70 beats per minute at rest, to over 100 beats per minute during exercise.

Punnett square A graphic method used in **genetics** to calculate the results of all possible **fertilizations** and hence the **genotypes** and **phenotypes** of **progeny**. In a Punnett square, the symbols used to represent one of the parent's **gamete** genotypes are written along the top and those of the other parent down the side. The permutations possible during fertilization are worked out by matching male and female gametes. See **monohybrid inheritance**, **backcross**, **incomplete dominance**.

pupa A stage in the **life history** of some insects between **larva** and **imago**, during which a radical change in form occurs. See **metamorphosis**.

pure-breeding Possessing an inherited trait, controlled by a **homozygous** pair of **alleles**, which in successive self-crosses reappears generation after generation. See **monohybrid inheritance**.

pyloric sphincter The ring of muscles which opens and closes the entry to the **stomach**.

pyramid of biomass A diagram illustrating that organisms at the end of a **food chain** have a smaller total mass (**biomass**) than those at the beginning of the chain. See **pyramid of numbers**.

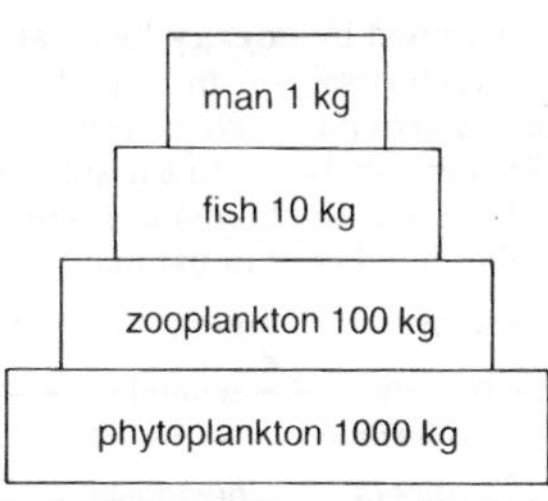

pyramid of biomass The pyramid of biomass in the Antarctic ecosystem.

pyramid of numbers A diagram illustrating the relationship between members of a **food chain**, showing that the organisms at the end of the chain are usually fewer in number, or to be more exact, have a smaller total mass (**biomass**) than those at the beginning of the chain.

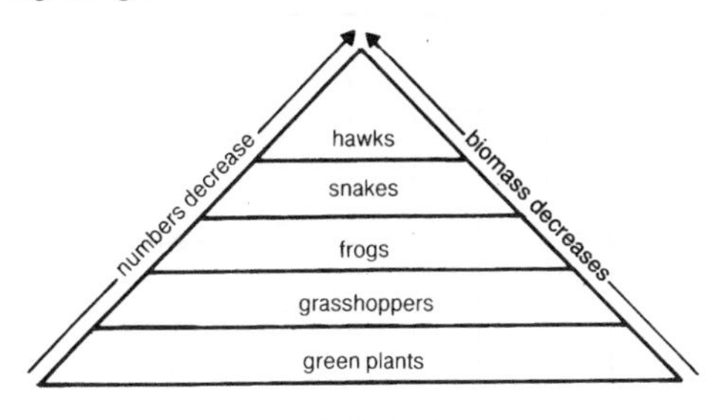

pyramid of numbers

The decrease is caused by **energy** losses at each link in the chain, i.e. each organism in a food chain uses up energy in various activities such as heat production and movement. This energy is lost to the subsequent organisms in the chain and so the reduced energy can only support a smaller number of individuals.

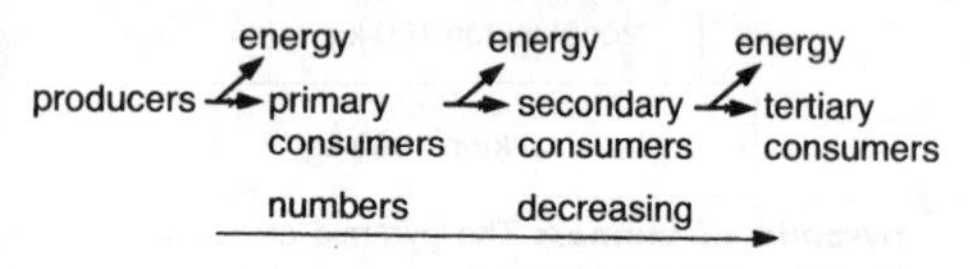

pyramid of numbers Energy loss.

quadrat A square frame used to sample plants (and stationary animals, e.g. barnacles). A quadrat marks off a small area so that the **species** present can be identified and counted. This sample gives an estimate of the

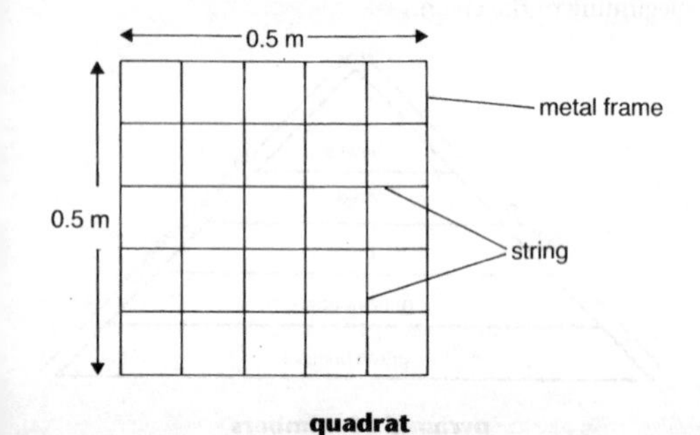

quadrat

numbers and types of species in the whole area. The quadrat must be placed randomly to obtain a representative sample and a number of samples taken.

radicle The embryonic **root** of **seed** plants which is the first structure to emerge from the seed during **germination**. See **germination**.

radius The **anterior** of the two **bones** of the lower region of the **tetrapod** forelimb. In humans, it is the shorter of the two bones of the forearm. See **endoskeleton**.

receptor or **sense organ** A specialized tissue in an animal which detects **stimuli** from the **environment** and which, by sending **nerve impulses** through the **nervous system**, causes **responses** to be made. See **sensitivity**.

recessive One of a pair of **alleles** which is only expressed in a **homozygous phenotype**. It is the converse of **dominant**. See **monohybrid inheritance**, **backcross**, **incomplete dominance**.

rectum The terminal part of the vertebrate **intestine** in which **faeces** are stored prior to expulsion via the **anus** or **cloaca**. See **digestion**.

red blood cell, **red blood corpuscle** or **erythrocyte** The most numerous **cell** of vertebrate **blood**, responsible for transporting oxygen from the **lungs** to the **tissues**. In humans, red blood cells are made in **bone** marrow and are biconcave discs, without **nuclei**.

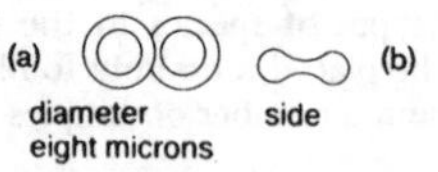

red blood cell Shape as seen (a) from above and (b) from the side.

Red blood cells contain **haemoglobin** which combines with oxygen as blood passes through the lungs, forming a compound called *oxyhaemoglobin*. At the tissues, this unstable compound breaks down, thus releasing oxygen to the cells.
A shortage of red blood cells is called *anaemia*.

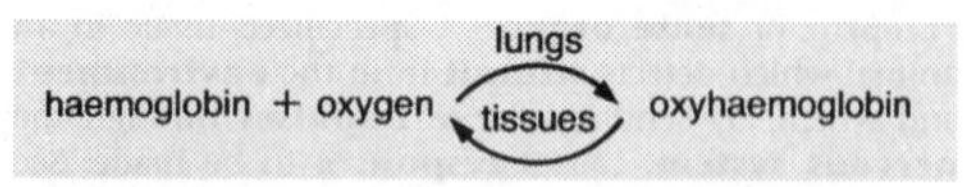

red blood cell Haemoglobin delivers oxygen to the tissues.

red blood corpuscle See **red blood cell**.

reflex action A rapid involuntary **response** to a **stimulus**, occurring in most animals and in vertebrates, mediated by the **spinal cord**. Reflex actions can be important in protecting animals from injury, by for instance withdrawing a limb from a hot object.
The structures involved and the **nerve impulses** responsible for reflex actions constitute a *reflex arc* which is set up when a nerve impulse is initiated at a receptor. The impulse is transmitted along a sensory **neurone** to the spinal cord where it crosses a **synapse** to a motor neurone.

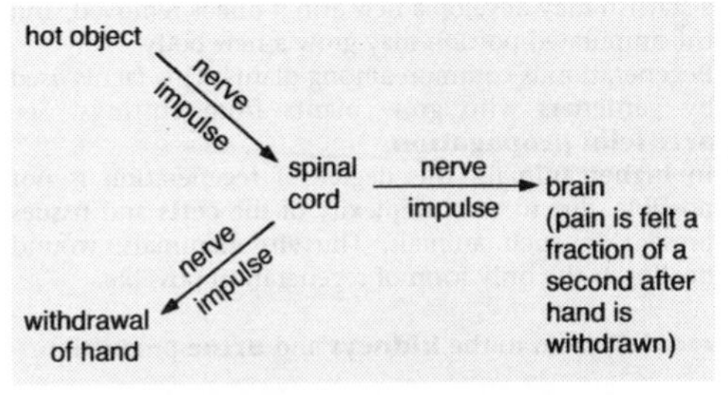

reflex action Path of nerve impulses.

When a reflex arc operates, nerve impulses are also sent from the spinal cord to the **brain**. Thus, although the response is initiated by the spinal cord, it is through the brain that the animal is aware of what has happened.

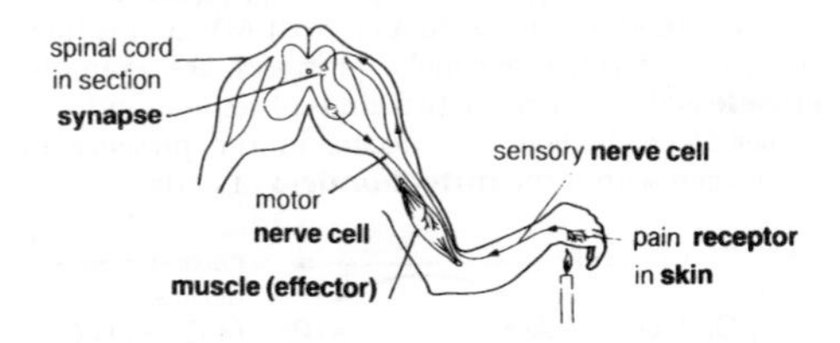

reflex action The reflex arc.

regeneration The regrowth by an organism of **tissues** or **organs** which have been damaged or removed. In certain lower animals whole new individuals can develop from portions of a damaged adult. For example,

a starfish may develop a new arm if one is removed, and the amputated portion may grow a new body.
Regeneration is common among plants. This fact is used by gardeners who grow plants from cuttings. See **artificial propagation**.
In higher animals, this degree of regeneration is not possible, due to the complexity of the **cells** and tissues present in such animals. Thus in mammals, wound healing is the only form of regeneration possible.

renal Related to the **kidneys** and **urine** production.

reproduction The process by which a new organism is produced from one or a pair of parent organisms. See **asexual reproduction, sexual reproduction**.

respiration The reactions by which organisms release the chemical **energy** of food, for example, **glucose**. The energy is used to synthesize **ATP** from ADP and is then available for other metabolic processes, for example, **muscle** action. There are two kinds of respiration:
(a) *aerobic respiration*: this occurs in the presence of oxygen within the **mitochondria** of **cells**;

glucose + oxygen	⟶		carbon dioxide + water
$C_6H_{12}O_6 + 6O_2$	ADP	ATP	$6CO_2 + 6H_2O$

respiration Aerobic respiration.

(b) *anaerobic respiration*: this occurs in the absence of oxygen within the **cytoplasm** of **cells**, and provides a lower ATP yield than aerobic respiration.

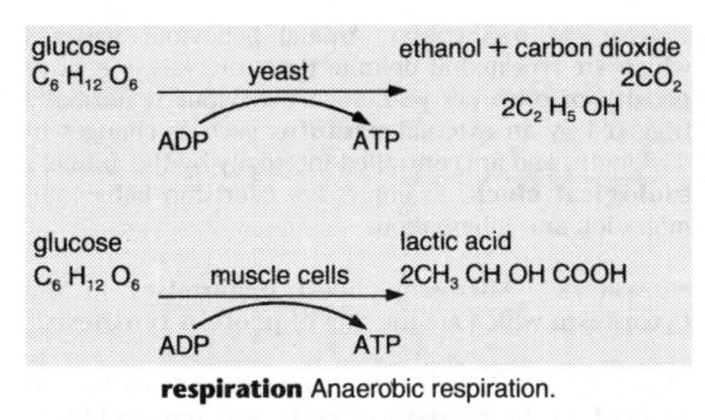

respiration Anaerobic respiration.

response Any change in an organism made in reaction to a **stimulus**. See **sensitivity**.

retina Light-sensitive **tissue** lining the interior of the vertebrate **eye**, and consisting of two types of **cell**: **rods** and **cones**.

rhizome An **organ** of **vegetative reproduction** in flowering plants, consisting of a horizontal underground stem growing from a parent plant. The tip of the rhizome is a bud from which grows a new plant. Plants which have rhizomes include iris and many types of grass.

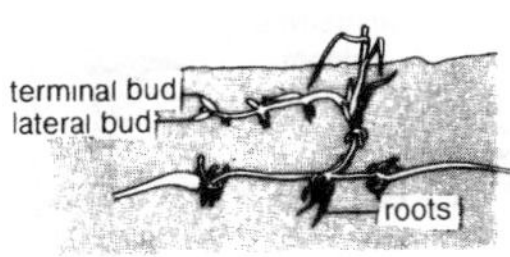

rhizome A grass rhizome.

rhythmical behaviour Animal behaviour patterns which are repeated at definite time intervals, e.g. once per day or once per year. Such behaviour is normally triggered by an external **stimulus** such as changes in daylength, and are controlled internally by the animal's **biological clock**. Examples are courtship behaviour, migration and hibernation.

ribosomes Microscopic **cell organelles** in the **cytoplasm** which are the sites of **protein synthesis**.

RNA (ribose nucleic acid) **Nucleic acid** synthesized by **DNA** in the **nucleus** of **cells**, and responsible for carrying the **genetic code** from the nucleus into the **cytoplasm** where the synthesis of **proteins** occurs. RNA differs from DNA in the following ways:
(a) RNA is a single polynucleotide chain;
(b) the sugar group is ribose;
(c) thymine is replaced by uracil.

rod A light-sensitive **neurone** in the **retina** of the vertebrate **eye** which can function in dim light.

root The part of a flowering plant that normally grows

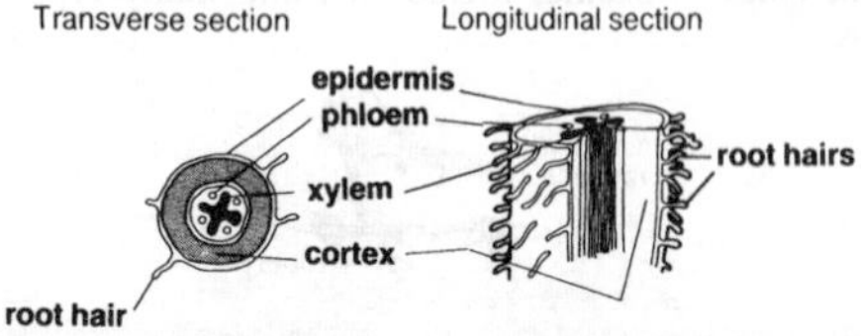

Transverse section
Longitudinal section
epidermis
phloem
xylem
cortex
root hairs
root hair

root The structure of a dicotyledon root.

down into the **soil**. Its functions are:
(a) absorption of **water** and **mineral salts** from soil;
(b) the anchoring of the plant in the soil;
(c) the storage of food, in some plants, e.g. turnip.

root cap A cap-shaped layer of **cells**, covering the apex of the growing root tip and protecting it as the root grows through the **soil**.

root cap Longitudinal section through root.

root hairs Tubular projections from **root epidermis cells**, the **nucleus** usually passing into the hair. Root hairs enormously increase the surface area of the root, and are the principal absorbing tissue of the plant. **Water** enters root hairs from the **soil** by osmosis, while **mineral salts** are absorbed by **active transport**. The water and mineral salts then pass through the **cortex**

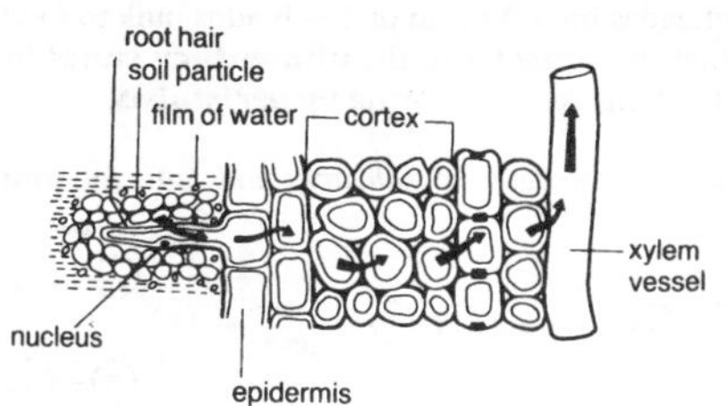

root hairs Water and mineral salts enter the plant via the root hairs.

cells and enter **xylem** vessels from where they are transported throughout the plant via the **transpiration stream**.

root nodules Swellings on the roots of leguminous plants (for example, clover, bean, pea). Root nodules contain **bacteria** of the **genus** *Rhizobium* which convert the nitrogen of **soil**-air into organic nitrogen compounds which can be used by the legumes. This is called **nitrogen fixation**. See **nitrogen** cycle.

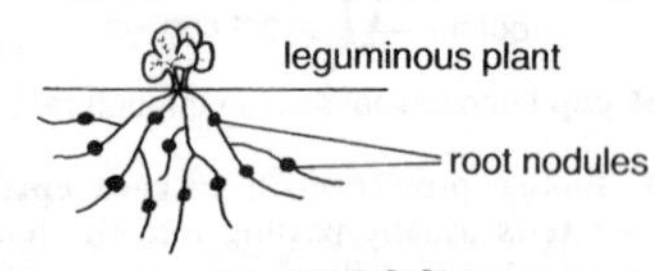

root nodules

roughage An important component of human **balanced diet**, consisting mainly of the **cellulose** in plant **cell** walls. Although indigestible by human beings, roughage is important in diet as it adds bulk to food and enables the **muscles** of the **alimentary canal** to grip the food and keep it moving by **peristalsis**.

saliva Fluid secreted by *salivary glands* into the **mouths**

starch
...G-G-G-G-G-G... salivary amylase → maltose
where G = glucose

saliva Amylase in saliva digests starch into maltose.

of many animals in order to moisten and lubricate food. In some mammals, including humans, saliva contains the **enzyme** *salivary amylase*.

saprophyte An organism that feeds on dead and decaying plants and animals, causing decomposition. Many **fungi** and **bacteria** are saprophytic and play an important role in recycling nutrients. See **carbon cycle**, **nitrogen cycle**.

scapula The **dorsal** part of the **tetrapod** shoulder girdle. In humans, it is the *shoulder blade*. See **endoskeleton**.

scientific method The procedures by which scientific investigations should be made. Scientific method involves the following steps:

(a) *observation*: an occurrence is seen to happen on more than one occasion. For example, it may be observed that **starch** in plant **seeds** apparently supplies **energy** during **germination**;

(b) *problem identification*: the observation is questioned. For example, how does starch which is a long chain **carbohydrate** become suitable as a respiratory **substrate**?

(c) **hypothesis**: the suggestion of a possible solution. For example, a **carbohydrase enzyme** within seeds degrades starch to **glucose**;

(d) *Experiment*: the hypothesis is tested. For, example, seed extract is added to starch, and a test for glucose is made;

(e) **theory**: the proposal of a solution to the problem based on experimental evidence; for example, that in plant seeds, a carbohydrase enzyme degrades starch to glucose which then acts as a respiratory substrate during germination.

All valid scientific investigations follow the guidelines of the scientific method and must include **control experiments** which are identical to the test experiment in all aspects except one. The control provides a standard with which the test experiment can be compared, by showing that any change occurring in the test experiment was due to the factor missing from the control and would not have happened anyway. For example, in the seed/starch experiment, a tube containing starch alone would be a suitable control.

sclerotic The external protective layer of the vertebrate eyeball. See **eye**.

secondary sexual characteristics Features (excluding **gonads** and associated structures) which distinguish between adult male and female animals. The development of such features, for example, lion's mane, stag's antlers, and in humans, breast development in females, facial hair in males, etc., is usually controlled by sex **hormones**. Compare **primary sexual characteristics**.

seed The structure that develops from an **ovule** after **fertilization** in flowering plants, and which grows into a new plant. Seeds are enclosed within a **fruit**.

Within the seed, the **embryo** becomes differentiated into an embryonic **shoot** bud (**plumule**) and **root** (**radicle**) and either one or two seed leaves (**cotyledons**).

See **fruit and seed dispersal, germination**.

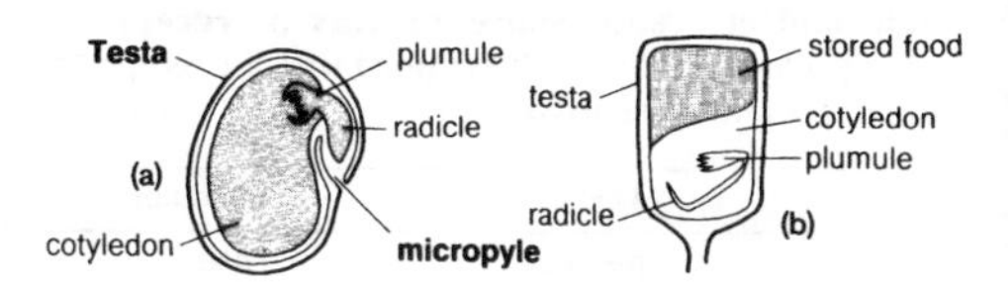

seed The structure of a seed: (a) broad bean (b) maize grain.

selective breeding See **artificial selection**.

selectively permeable membrane Membranes surrounding **cells** and bordering regions within cells. They are made up of orderly arrangements of **protein** and **fat molecules**. Certain small molecules may pass through pores in the membrane but larger ones are held back. For this reason the membrane is described as selectively permeable or *semipermeable*. See **cell**, **osmosis**.

semicircular canals Tubes within the vertebrate inner **ear**, which are important in maintaining balance. See **ear**.

sense organ See **receptor**, **sensitivity**.

sensitivity or **irritability** The ability of living organisms to respond to changes in environmental **stimuli**, such as heat, light, sound, etc. Sensitivity enables organisms to be aware of changes in their environment and thus to make appropriate **responses** to any changes that may occur. Certain parts of animals, for example, **eyes**, **ears**, **skin**, are sensitive to particular environmental stimuli and are called **sense organs** or **receptors**. Similarly, plant **tissues** such as **shoot** tips, are receptors, being important in **tropisms**.

Sense	Stimulus	Receptor
smell	chemicals	nose
taste	chemicals	mouth
touch	contact	skin
hearing	sound	ears
sight	light	eyes
balance	change of position	inner ear

sensitivity

As a result of stimuli from the environment, responses are initiated in specialized structures called **effectors**, for example, **muscles**. The responses made by an organism constitute its behaviour.

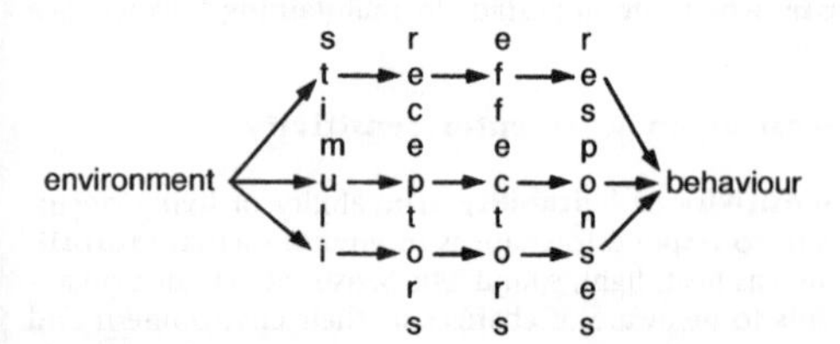

sensitivity Organisms respond to changes in their environment.

In mammals the receptors are specialized **cells** connected to the **brain** or **spinal cord** which together make up the **central nervous system** (CNS). In response to a stimulus, the receptors initiate a **nerve impulse** which is transmitted by **neurones** to the CNS, i.e. the receptor converts the **energy** of the stimulus into the *electrical energy* of the nerve impulse.
See **smell, taste**.

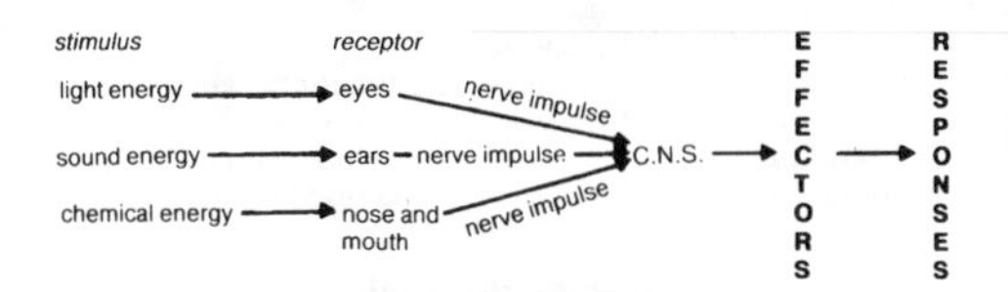

sensitivity A stimulated receptor sends a nerve impulse to the CNS which transmits a response.

sewage disposal The treatment of sewage in order to make it harmless. Sewage, which is mainly domestic waste, is first filtered to remove large particles, and then piped into tanks where solids settle out as *sludge*. The remaining liquid is aerated to encourage the growth of *aerobic bacteria* which feed on dissolved **organic compounds**. The **bacteria** are, in turn, eaten by larger organisms and when the dead organisms are allowed to settle, the remaining liquid can be safely discharged into a river. Sludge may also be digested by **microbe** action, with the production of **methane** gas (also called *biogas*) which is used as a fuel.

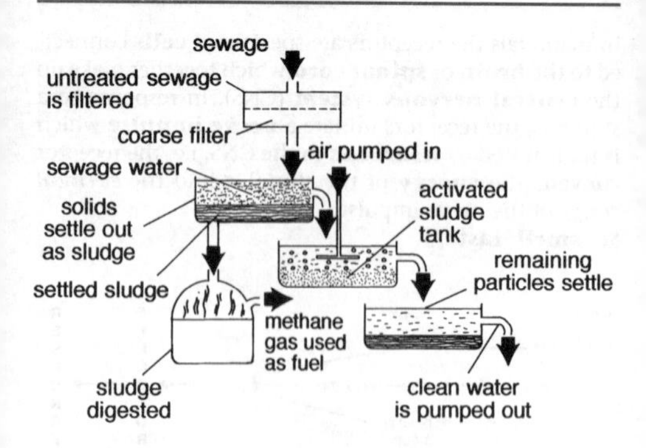

sewage disposal

sex chromosome Any chromosome that is involved in **sex determination**. In **diploid** human **cells** there are 46 chromosomes made up of 23 **homologous chromosome** pairs, one of the pairs being described as sex chromosomes. In the female, the two sex chromosomes are similar (*homogametic*) and are called X chromosomes. The female **genotype** is thus XX. In the male, one of the pair is distinctly smaller, and is called the Y chromosome. The male genotype is thus XY (*heterogametic*).
The male is not always the heterogametic sex. For example, in birds, the male is XX, and the female is XY, while in some insects the female is XX and the male is XO, the Y chromosome being absent.
The sex chromosomes, as well as determining sex, also contain **genes** controlling other traits, resulting in what is known as **sex linkage**.

sex determination The method by which the sex of a **zygote** is determined, the most common method being by **sex chromosomes**. Consider the **genotypes** of (a) a human male and (b) a human female:

(a) a Y-bearing sperm may fertilize an ovum giving a zygote genotype XY and **phenotype** male;

(b) an X-bearing sperm may fertilize an ovum giving a zygote genotype XX and phenotype female. Since half the sperms are X and half are Y, there is an equal chance of the zygote being male or female.

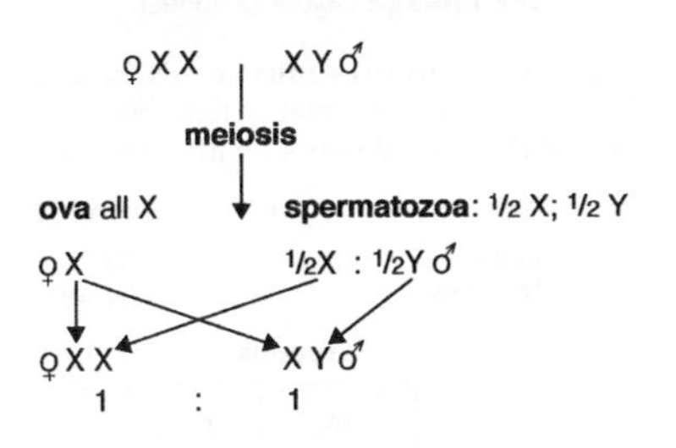

sex determination

sex linkage The presence on a **sex chromosome** of **genes** unconnected with sexuality, resulting in certain traits appearing in only one sex. In humans, most sex-linked genes are carried on X chromosomes, the Y chromosome being concerned mainly with sexuality.

Example:

The gene for colour blindness in humans is carried on the X chromosome. Normal vision is **dominant** to colour blindness

If N = normal and n = colour blind:

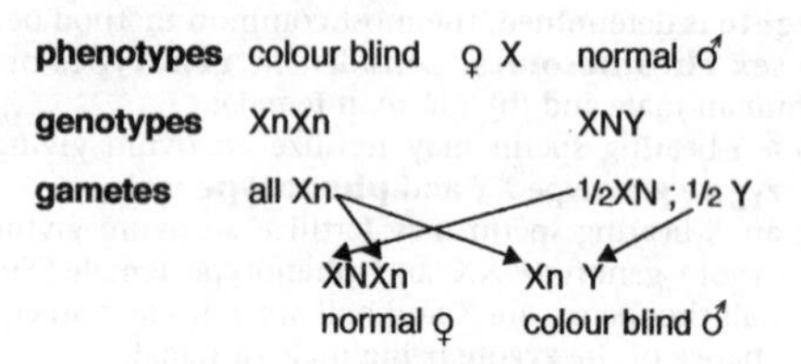

sex linkage Colour blindness.

In this case, the **heterozygous** ♀ XNXn is called a 'carrier' since she has normal vision but carries the **recessive allele**. Thus if crossed with a normal male:

	carrier ♀	×	normal ♂
	XNXn		XNY
gametes	$\frac{1}{2}$XN; $\frac{1}{2}$Xn		$\frac{1}{2}$XN; $\frac{1}{2}$Y

sperms

ova		XN	Y
	XN	XNXN	XNY
	Xn	XNXn	XnY

Progeny			
	genotypes	XNXN	XNXn
	phenotypes	normal ♀	carrier ♀
	genotypes	XNY	XnY
	phenotypes	normal ♂	colour blind ♂

sex linkage The inheritance of colour blindness from a carrier.

That is, the result from the **Punnett square** indicates a possibility that half the sons will be colour blind, and half the daughters will be carriers.
A more serious sex-linked trait is **haemophilia** (prolonged bleeding) but its transmission and inheritance is the same as above.

sexual reproduction **Reproduction** involving the joining or fusing of two sex cells (**gametes**) one from a male parent, and one from a female parent. Gametes are **haploid** and when they fuse (**fertilization**), the resulting composite cell (**zygote**) has the **diploid** number of **chromosomes**. After fertilization, the diploid zygote divides repeatedly, ultimately resulting in a new organism. The diagram shows this for humans, who have 46 chromosomes.
The offspring of sexual reproduction are genetically unique (except for identical twins) because, unlike the offspring of **asexual reproduction**, they obtain half their chromosomes from their male parent and half from their female parent. Thus each fertilization produces a new combination of chromosomes which in turn produce a new organism which will likewise produce gametes by **meiosis**.

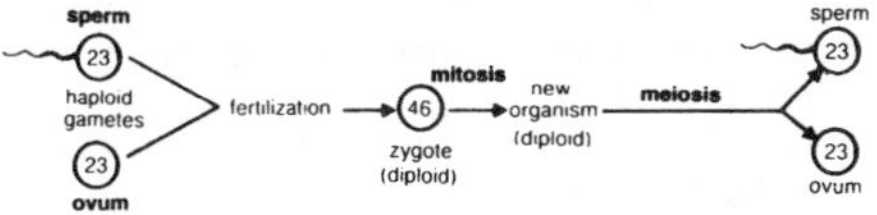

sexual reproduction Two haploid gametes fuse to form a diploid zygote.

shoot That part of a flowering plant which is above **soil** level, for example, **stem**, **leaves**, buds, **flowers**.

short sight or **myopia** A human eye defect, mainly caused by the distance from **lens** to **retina** being longer than normal. This results in distant objects being focused in front of the retina giving blurred vision. Short sight is corrected by wearing diverging (*concave*) lenses.

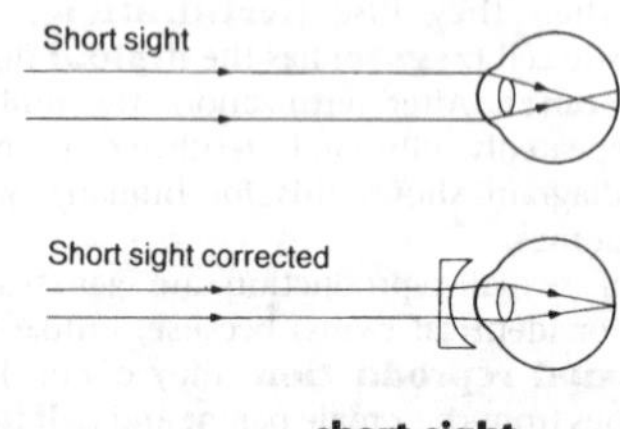

short sight

single-cell protein (SCP) A protein-rich food supplement produced by certain **microorganisms**, which is fed to domestic animals. Since microorganisms reproduce very rapidly, and are more efficient protein producers than other organisms, they can be grown, harvested and dried to produce SCP.
The figures below compare the protein produced per day by 1000 kg of **biomass**:

cattle 1kg
soya beans 100 kg
microorganisms 10^{15} kg.

skeleton The hard framework of an animal which supports and protects the internal organs. See **endoskeleton**, **exoskeleton**.

skin The layer of **epithelial cells**, **connective tissue** and associated structures, that covers most of the body of vertebrates. Mammalian skin consists of two main layers:

(a) the **epidermis**; this is the outer layer. It consists of
  (i) *cornified layer*: dead cells forming a tough protective outer coat;
  (ii) *granular layer*: living cells which ultimately form the cornified layer;
  (iii) *Malpighian layer*: actively dividing cells which produce new epidermis;

(b) the **dermis**; this is a thicker layer below the epidermis, containing **blood capillaries**, hair follicles, sweat glands and **receptor** cells sensitive to touch, heat, cold, pain, pressure.

Beneath the dermis, there is a layer of **fat** storage cells which also act as heat insulation

The functions of mammalian skin are:

(a) to protect against injury and **microorganism** entry;

(b) to reduce **water** loss by evaporation;

(c) to act as **receptor** for certain environmental **stimuli**;

(d) to play an important part in **temperature regulation** in **endotherms**.

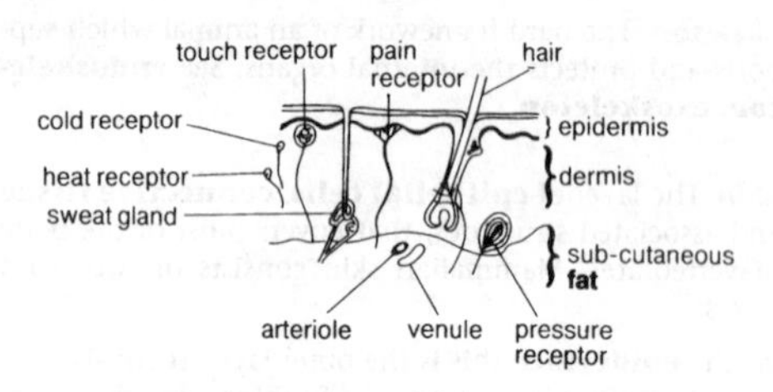

skin Section through mammalian skin.

small intestine The **anterior** region of the vertebrate **intestine**. In man, it consists of the **duodenum** (about thirty centimetres in length) and the **ileum** (about seven metres in length). The duodenum receives food from the **stomach**. See **digestion**.

smell The ability of animals to detect odours. In humans, the **receptor cells** involved are in the nasal cavity, and are sensitive to chemical **stimuli**. See **sensitivity**.

smog A harmful side-effect of smoke **pollution** caused by particles of smoke sticking to droplets of water in the atmosphere forming a thick mist:

smoke+fog→smog

Such a smog is thought to have resulted in 4000 deaths from respiratory disease in London in 1952. Since the Clean Air Act of 1956, smog has been eliminated in Britain, but is still prevalent in other industrialized countries.

smoking The drawing into the **lungs** of tobacco smoke, usually from cigarettes. Smoking is habit-forming and can cause lung cancer, bronchitis, and other serious diseases.

smooth muscle See **involuntary muscle**.

soil The weathered layer of the Earth's crust intermingled with living organisms and the products of their decay.

The components of soil are:
(a) inorganic particles (weathered rock fragments);
(b) **water**;
(c) **humus**;
(d) air;
(e) **mineral** salts;
(f) **microorganisms**;
(g) other organisms (for example, earthworms).

Soil is important because
(a) it is a **habitat** for a wide variety of organisms;
(b) it provides plants with water and mineral salts;
(c) decomposition of dead organisms in soil releases minerals which can be used by other living organisms.

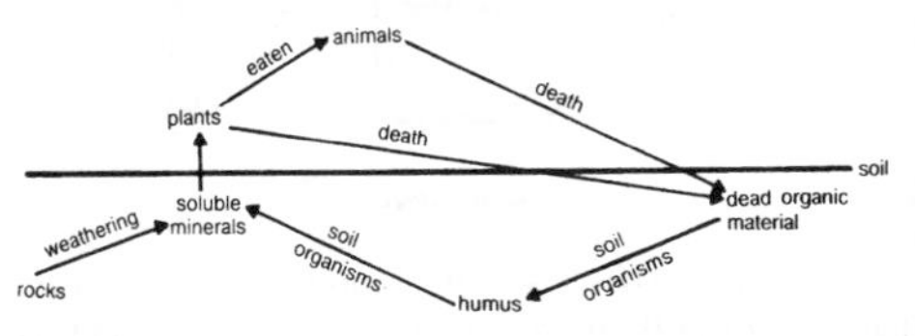

soil Dead organic material is broken down in the soil.

soil depletion The loss of **mineral salts** from **soil** when a crop is harvested, which like **soil erosion** may render the soil infertile. Soil depletion can be prevented by
(a) **crop rotation**;
(b) the addition of **fertilizers**.

soil erosion The loss of soil due to the agricultural practices associated with crop growing, such as repeated ploughing, deforestation, etc., which make the top-soil, which is rich in **humus** and minerals, less stable and more vulnerable to the effects of wind and rain. See **soil depletion**.

soil sieve A device used to separate the four types of inorganic particles in **soil**. The proportion of these particles in a soil can be measured by passing a weighed sample of dried soil through sieves of varying mesh size which separate the particles by size. The separated particles are weighed and their percentage of the complete sample can be calculated. See **soil texture**.

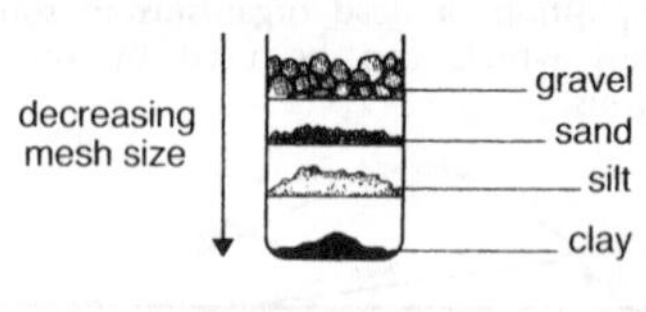

soil sieve

soil texture The types and proportions of inorganic particles in **soil**, of which four types are recognised, based on size.

clay salt sand gravel

increasing particle size

soil texture

Soil texture has important effects on soil properties such as water retention and aeration. See **soil types**.

soil thermometer A Celsius thermometer adapted for measuring **soil** temperature.

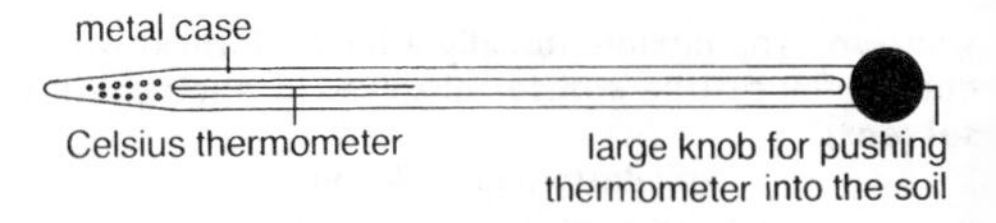

soil thermometer

soil types The numerous types of **soil** that exist. A simple classification recognizes three distinct types:

(a) *sandy (light) soil* has a high proportion of the larger inorganic particles, and hence larger air spaces. Thus, sandy soil is well aerated and has good drainage but tends to lose **mineral salts** which are washed downwards (**leaching**);

(b) *clay (heavy) soil* has a high proportion of small particles, which means that it retains minerals, but is poorly aerated and can become waterlogged;

(c) *loam soil* is the most fertile soil, consisting of a balance of particle types and a good **humus** content. Soils of this type are well-aerated and drain freely, but still retain water and minerals.

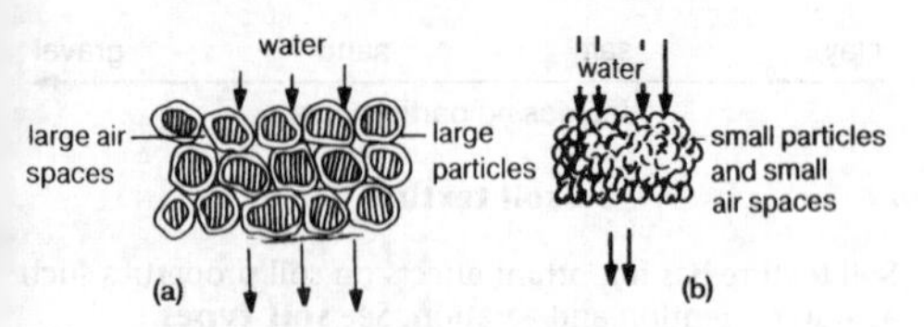

soil types (a) Sandy soil. (b) Clay soil.

solute The substance, in a **solution**, which dissolves in the **solvent**.

solution The mixture (usually a liquid) formed when one substance (the **solute**) dissolves in another (the **solvent**),

solute+solvent→solution
e.g. sugar+water→sugar solution

solvent The liquid, in a **solution**, in which a **solute** dissolves.

species A unit used in the **classification** of living organisms. It is a group of organisms which share the same general physical characteristics and which can mate and produce fertile offspring. For example, all dogs, despite variation in shape, size etc., are of the same species, but horses and donkeys are separate species within the same **genus**.

spermatozoon (*pl.* **spermatazoa**) or **sperm** The small motile male **gamete** formed in animal **testes**, and usually having a **flagellum**. Sperms are released from the male in order to fertilize the female gamete. See **fertilization**, **meiosis**.

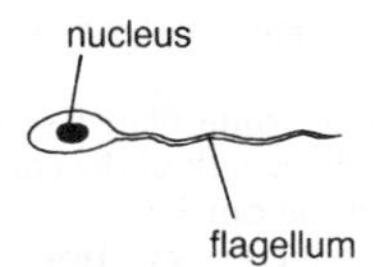

spermatazoon

sphincter A ring of **muscle** around tubular **organs**, which by contracting, can narrow or close the passage within the organ.
Examples are the anal sphincter (at the **anus**), and the **pyloric sphincter**. See **digestion**.

spinal cord That part of the vertebrate **central nervous system** which is enclosed within and protected by the **vertical column** (backbone).
The spinal cord is a cylindrical mass of **neurones** which connect with the **brain** and also with other parts of the body via *spinal nerves*. The spinal cord consists of three regions:

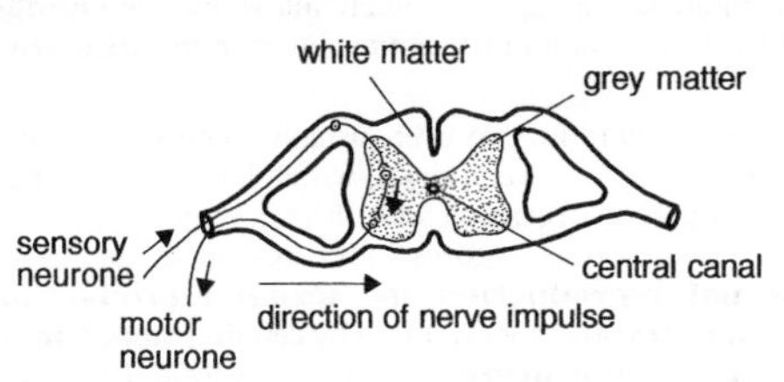

spinal cord Section showing the three regions.

(a) an inner layer of grey matter consisting of neurone cell bodies,
(b) an outer layer of white matter consisting of nerve fibres running the length of the cord;
(c) a fluid-filled central canal.

The spinal cord conducts **nerve impulses** to and from the brain and is also involved in **reflex actions**.

spiracle One of many pores in the **cuticle** of insects, connecting the **tracheae** with the atmosphere. See **gas exchange**.

spleen An **organ** in the **abdomen**, near the **stomach**, in most vertebrates. It produces **white blood cells**, destroys worn out **red blood cells**, and filters foreign bodies from the **blood**.

spongy mesophyll **Tissue** in a **leaf** situated between the **palisade mesophyll** and the lower **epidermis**. Spongy mesophyll **cells** are loosely packed, being separated by air spaces which allow **gas exchange** between the leaf and the atmosphere via the **stomata**.

spore A reproductive unit, usually microscopic, consisting of one or several **cells**, which becomes detached from a parent organism and ultimately gives rise to a new individual. Spores are involved in both **asexual reproduction** and **sexual reproduction** (as **gametes**) and are produced by certain plants, **fungi**, **bacteria** and **protozoa**. Some spores form a resistant resting stage of a **life history** while others allow rapid colonization of new **habitats**.

spore Spore release in the bread mould *Mucor*.

stamen The male part of a **flower** in which pollen grains are produced. Each stamen consists of a stalk (*filament*) bearing an **anther**.

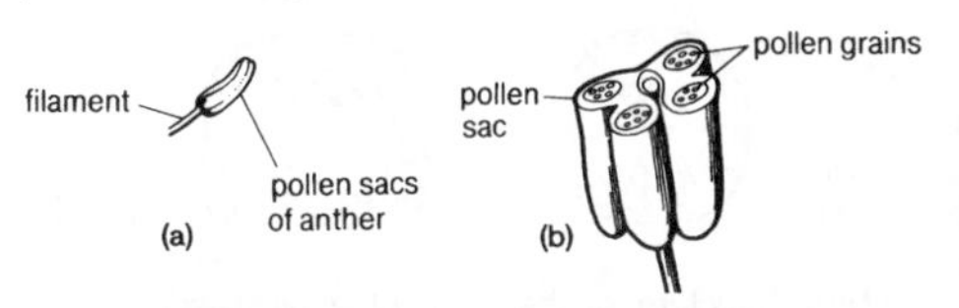

stamen (a) Stamen. (b) Anther cut open.

starch A **polysaccharide carbohydrate** which consists of chains of **glucose** units and which is important as an **energy** store in plants. Starch is synthesized during **photosynthesis** and is readily converted to glucose by **carbohydrase enzymes**. See **polysaccharides**.

stem That part of a flowering plant that bears the buds, **leaves** and **flowers**. Its functions are:

(a) the transport of **water**, **mineral salts** and **carbohydrate**;

(b) the raising of the leaves above the soil for maximum air and light;

(c) the raising of the **flowers**, thus aiding **pollination**;

(d) (in green stems) **photosynthesis**.

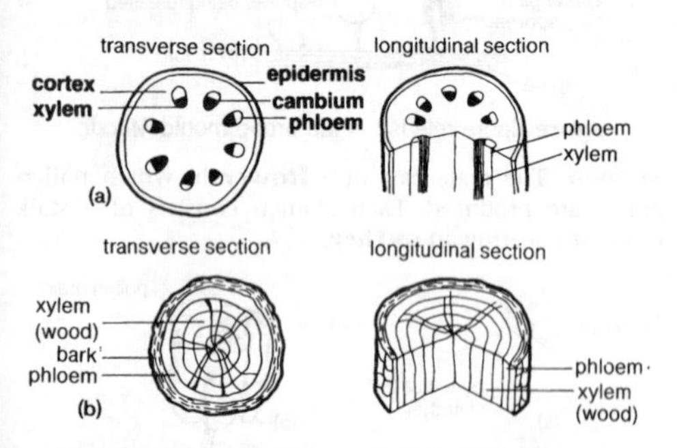

stem Structure of the stem of a dicotyledonous plant. Transverse and longitudinal sections of (a) a young stem and (b) an older stem.

sterilization **1.** A procedure to make an organism incapable of **reproduction**.
2. A procedure to make materials free of **microorganisms**. See **autoclave**.

sternum (or **breastbone**) A **bone** in the middle of the **ventral** side of the **thorax** of **tetrapods**, to which most of the ventral ribs are attached. See **endoskeleton**.

stigma A sticky structure in a flower which traps incoming pollen during **pollination**. See **fertilization**.

stimulus Any change in the **environment** of an organism which may provoke a **response** in the organism. See **sensitivity**.

stolon An **organ** of **vegetative reproduction** in flowering plants consisting of a horizontal **stem** growing from a bud on the parent organism's stem. Stolons grow above the **soil** and eventually the tip becomes established in the soil and develops into an independent plant.

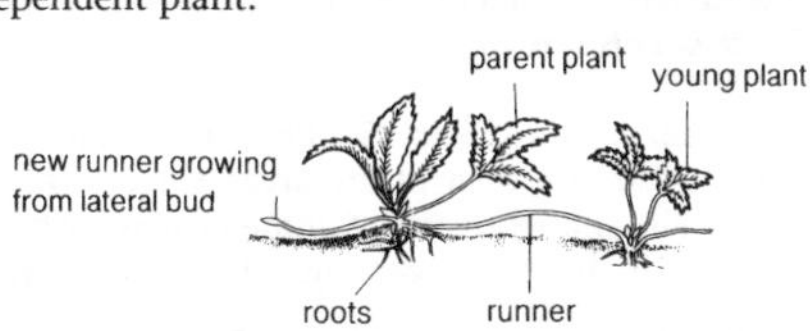

stolon A strawberry runner is an example of a stolon.

stoma (*pl.* **stomata**) One of many small pores in the **epidermis** of plants, particularly **leaves**. The evaporation of water during **transpiration**, and **gas exchange**, occur via the stomata. See **guard cells**.

stomach The muscular sac in the **anterior** region of the **alimentary canal**.
In vertebrates, food is passed to the stomach by **peristalsis** via the **oesophagus**.
In the stomach, food is mechanically churned by the peristaltic action of the walls, and **protein digestion** is initiated. See **pepsin**.
In **herbivores**, the stomach has several chambers for **cellulose** digestion.

From the stomach, food is passed into the **small intestine** through the **pyloric sphincter**.

striated muscle See **voluntary muscle**.

substrate A substance which is acted upon by an **enzyme**.

sugars **Water**-soluble, sweet-tasting crystalline **carbohydrates**, which include the **monosaccharides** and the **disaccharides**.

surface area/volume ratio The important relationship between the surface area of an organism or structure (such as a cell) and its overall volume or mass, which is significant to living organisms in several ways.

$$\frac{\text{surface area}}{\text{volume}} \quad \text{or} \quad \frac{\text{surface area}}{\text{mass}}$$

It is difficult to measure the surface area and volume of a plant or animal, but by using cubes as model organisms, the importance of the ratio can be seen. As the object becomes larger, its surface area becomes smaller relative to its volume. In living organisms this ratio has special significance in terms of heat and **water** loss.

(a) *Surface area/volume, and heat loss*: heat is lost more rapidly from small animals across their relatively large surface area, with the following consequences:
(i) small mammals such as mice eat relatively more food than larger mammals in order to generate **energy** to replace their high heat losses;
(ii) very small birds and mammals are restricted to

warm climates;

(iii) birds and mammals in cold **habitats** are usually larger than the same **species** living in warm climates.

(b) *Surface area/volume, and water loss*: relative to their volume, small organisms have a larger evaporating surface and thus a greater tendency to lose water. This is important because many animals and plants have problems controlling water balance, so the smaller the organism the greater the problem.

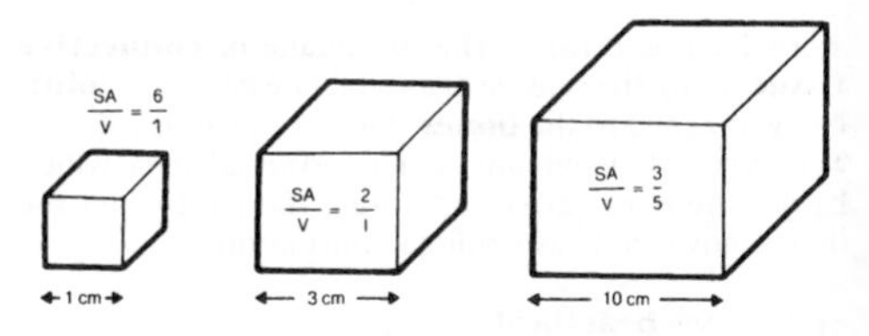

surface area/volume ratio

suspensory ligaments Structures holding the **lens** in place in the vertebrate **eye**. See **accommodation**.

symbiosis A relationship between organisms of different **species** for the purpose of nutrition. Examples of symbiosis include **parasitism**, **mutualism** and **commensalism**, although the term is sometimes restricted to mutualism.

synapse A microscopic gap between the **axon** of one **neurone** and the *dendrites* of another, across which a **nerve impulse** must pass. Nerve impulses arriving at a synapse cause **diffusion** of a chemical substance which crosses the gap to initiate nerve impulses in the next neurone.

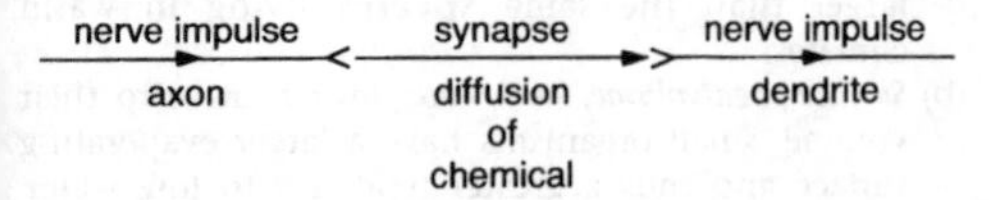

synapse A chemical substance diffuses across the synapse and initiates nerve impulses in the next nerve cell.

synovial membrane The membrane of **connective tissue** lining the capsule of a vertebrate moveable **joint**, being attached to the **bones** at either side of the joint. The synovial membrane secretes *synovial fluid* which bathes the joint cavity, lubricating the joint when the bones move, and cushioning against jarring.

systole See **heartbeat**.

taste The ability of animals to detect flavours. In humans, the **receptor cells** involved are *taste buds* which are sensitive to chemical **stimuli**, and are restricted to the mouth, particularly the tongue. There are four types of taste bud, sensitive to sweetness, sourness, saltiness and bitterness. See **sensitivity**.

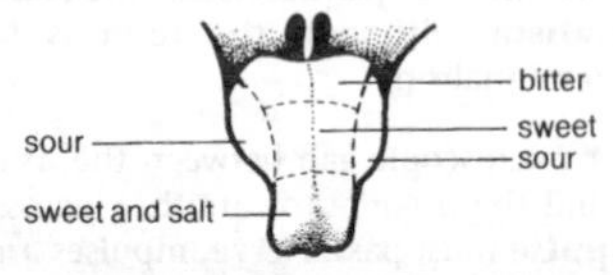

taste A taste-map of the tongue.

taxis (*pl.* **taxes**) A locomotory movement of a simple organism or a **cell** in response to an **environmental stimulus** such as light.

Such movements show a relationship to the direction of the stimulus, the movement either being towards (positive) or away (negative) from the source of the stimulus. Taxes are named by adding a prefix which relates to the stimulus. Thus a taxis relative to light is a *phototaxis*. Examples:

(a) Paramecium is *negatively geotactic*, i.e. it swims away from gravity;
(b) Fruit flies are *positively phototactic*, i.e. they move towards light;
(c) Many **spermatozoa** are *positively chemotactic*, i.e. they move towards chemical substances released by **ova**.

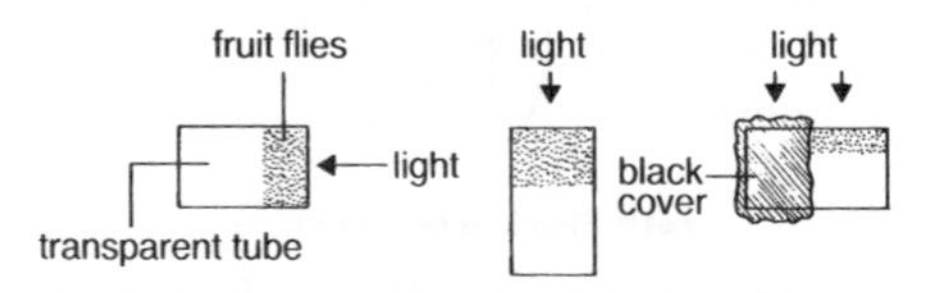

taxis Positive phototaxis in fruit flies.

teeth Structures within the mouth of vertebrates, used for biting, tearing, and crushing food before it is swallowed. Teeth consist of the following materials:

(a) *enamel*: a hard substance covering the exposed surface of the tooth (the *crown*). It contains calcium phosphate and provides an efficient biting surface.
(b) *dentine*: a substance similar to **bone**, forming the inner part of the tooth.

(c) *pulp*: soft **tissue** in the centre of the tooth containing **blood capillaries** which supply food and oxygen, and sensory **neurones** which register pain if the tooth is damaged.

(d) *root*: the part of the tooth within the gum, which is embedded in the jawbone by a substance called *cement*.

The types of teeth are **incisors**, **canines**, **premolars**, **molars** and **carnassials**. See **dental formula**, **dentition**, **carnivore**, **herbivore**, **omnivore**.

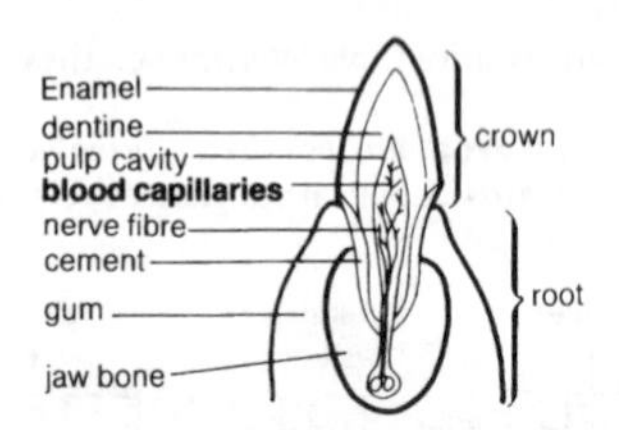

teeth Structure of a tooth.

temperature regulation The mechanisms, in **endotherms**, involved in maintaining body temperature within a narrow range (for example, in humans, close to 37 °C) so that the normal reactions of **metabolism** can take place.

Some of the temperature regulators employed by birds and mammals are outlined below:

(a) **fat** under the **skin** (*subcutaneous fat*) acts as an insulator;

(b) hair in mammals, and feathers in birds, trap air which is a good insulator;

(c) evaporation of sweat from the skin surface of some mammals has a cooling effect;
(d) *vasoconstriction*: superficial **blood vessels** constrict in response to cold, diverting blood away from the skin surface, and thus reducing heat loss;
(e) *vasodilation*: superficial blood vessels dilate in response to heat, bringing blood to the skin surface, from which heat can be lost to the atmosphere.

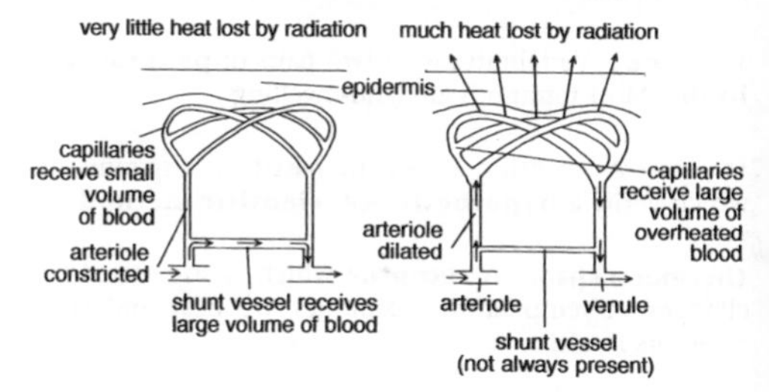

temperature regulation Vasoconstriction and vasodilation.

tendon A band of **connective tissue** by which **muscles** are attached to **bones**.

tensile strength The ability to withstand a certain amount of bending before breaking. See **bone**, **collagen**.

testa The outer protective coat of a **seed** formed from the **integuments** of the **ovule**, after **fertilization**. The testa is usually hard and dry and protects the seed from **microorganisms** and insects.

testcross See **backcross**.

testis The principal reproductive organ in male animals, which produces **spermatozoa**. In vertebrates, the paired testes also produce sex **hormones**. See **fertilization**.

tetrapods Vertebrates with two pairs of **pentadactyl limbs**. Most tetrapods are land-dwelling.

theory A scientific statement based on experiments which verify a **hypothesis**. See **scientific method**.

thermoreceptor A **receptor** which is stimulated by changes in temperature. Examples are heat and cold receptors in **skin**.

thorax **1.** (in vertebrates) The part of the body containing **heart** and **lungs** (*chest cavity*). In mammals, it is separated from the **abdomen** by the **diaphragm**.
2. (in insects) The part of the body **anterior** to the abdomen.

thyroid gland An **endocrine gland** in the neck region of vertebrates. When stimulated by *thyroid-stimulating hormone (TSH)* from the **pituitary gland** it produces the **hormone** *thyroxine* which controls the rate of growth and development in young animals. For example, in tadpoles, thyroxine stimulates **metamorphosis**.

thyroid-stimulating hormone (TSH) See **thyroid gland**.

thyroxine See **thyroid gland**.

tibia **1.** One of the segments of the insect leg.
2. The **anterior** of the two **bones** in the lower hindlimb of **tetrapods**. In humans, it is the shinbone. See **endoskeleton**.

tissue A group of similar **cells** specialized to perform a specific function in **multicellular** organisms. Examples are **muscle**, **xylem**.

tissue fluid See **lymph**.

toxin A substance secreted by, for example, **bacteria**, which is harmful to the organism within which the bacteria are living. See **antibodies**.

trachea **1.** (in land vertebrates) The windpipe leading from the **larynx** and carrying air to the **lungs** where it divides into the **bronchi**. The trachea is supported by **cartilage** rings and has a ciliated **epithelium** that secretes **mucus** which traps dust and **microorganisms**.
2. (in insects) One of the air tubes in a branching system through which air diffuses into the **tissues** via the **spiracles**. See **gas exchange**.

transect A line marked off in an area, to study the types of **species** in that area, by sampling the organisms at different points along the line. Measurements of

abiotic factors, for example, light, **soil pH** etc., may also be made along the line to discover any relationship between the distribution of particular species and these factors. See **quadrat**.

10 m string pegged and marked off in metres
soil level
quadrat
light meter
pH meter
soil thermometer

transect

transfer RNA See **RNA**, **protein synthesis**.

translocation The transport and circulation of materials within plants,

(a) of water and **mineral salts** in **xylem** vessels via the **transpiration stream**;
(b) of **carbohydrate** produced by **photosynthesis** and conducted through the plant in **phloem** sieve tubes.

transpiration The evaporation of **water** vapour from plant **leaves** via the **stomata**.

transpiration rate The rate at which **water** vapour is lost from a **leaf** to the outside atmosphere. **Transpiration** is affected by several environmental factors

(a) *temperature*: increased temperature increases water evaporation and thus increases transpiration;

(b) *humidity* (water content of air): increased humidity causes the atmosphere to become saturated with water, thus reducing transpiration;
(c) *wind*: increased air movements accelerate transpiration by preventing the atmosphere around **stomata** from becoming saturated with water.

Thus the transpiration rate will be greatest in warm, dry, windy conditions. If the rate of water loss by transpiration exceeds the rate of water uptake, **wilting** may occur.

transpiration stream The flow of **water** through a plant resulting from **transpiration**. Water evaporates through the **stomata**, causing more water to be drawn by **osmosis** from adjacent **leaf cells** (**spongy mesophyll** cells).

The osmotic forces thus set up eventually cause water to be withdrawn from **xylem** vessels in the leaf, resulting in water being pulled through the xylem vessels from the

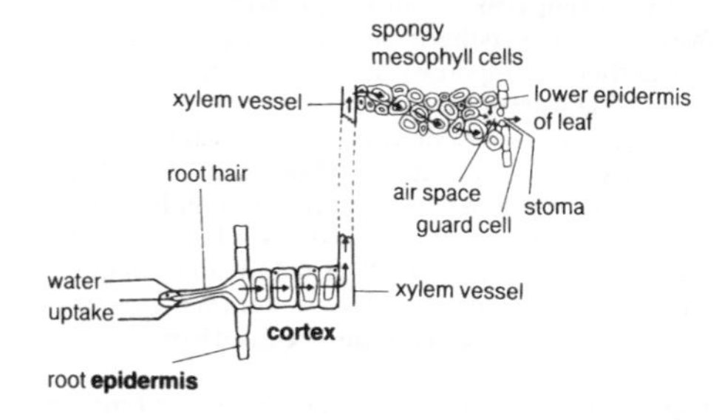

transpiration stream

stem and **roots**, i.e. water evaporation from leaves causes the flow of water (and **mineral salts**) throughout the plant.

trophic level The level in a **food chain** at which a group of organisms occurs. Green plants (**food producers**) are at the lowest level and tertiary consumers (**predators**) at the highest level.

tropism A plant **growth** movement which occurs in response to a **stimulus** such as light.
Such movements are related to the direction of the stimulus, the plant **organ** involved growing either towards or away from it. Tropisms are named by adding a prefix which relates to the stimulus.
Examples of tropisms are:
(a) **geotropism**: response to gravity;
(b) **phototropism**: response to light;
(c) **chemotropism**: response to chemicals;
(d) **hydrotropism**: response to **water**.
Tropisms can be either positive or negative depending on whether the response is towards the stimulus or away from it. See **nastic movement**.
Tropisms are important because they cause plants to grow in such a way that they obtain maximum benefit from the **environment**, in terms of water, light, etc.
Tropisms are caused by an **auxin** which accelerates growth by stimulating **cell division** and elongation. Uneven distribution of auxin causes uneven growth and leads to bending. See **tropism mechanism**.

tropism mechanism The method by which **tropisms** are controlled by plant **auxins**. The mechanism can be explained by considering experiments done on plant **growth** and **phototropism** using growing **shoots**.

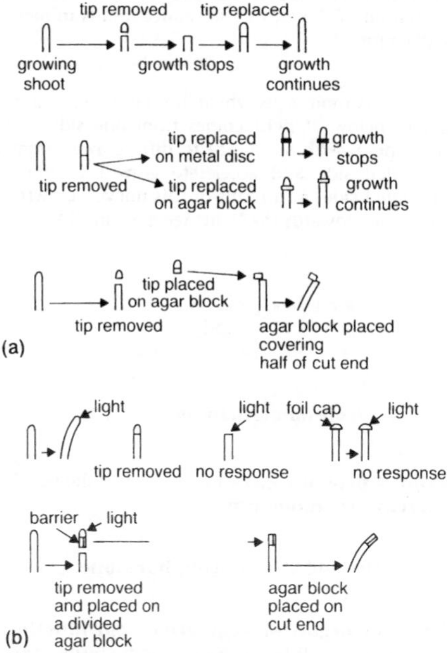

tropism mechanism (a) Shoot tips contain a growth substance. (b) Its action is affected by light.

A growth substance is produced in shoot tips. This substance diffuses downwards and accelerates growth by stimulating **cell division** and elongation. Uneven distribution of the growth substance results in bending. See diagram (a).

Shoot tips contain cells which are sensitive to light. See diagram below. If light comes from one side only, it causes the growth substance to diffuse away from the illuminated side and accelerate growth at the non-illuminated side. As a result of this uneven growth, the shoot bends towards the light. See diagram (b).

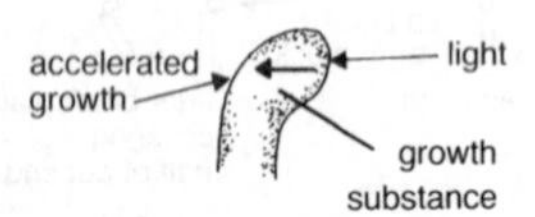

tropism mechanism A shoot tip.

trypsin A **protease enzyme** secreted by the vertebrate **pancreas**. See **duodenum**.

TSH See **thyroid-stimulating hormone**.

tuber An **organ** of **vegetative reproduction** in flowering plants. Tubers can form from **stems** or **roots**. They consist of a food store and buds from which develop new plants. See diagram.

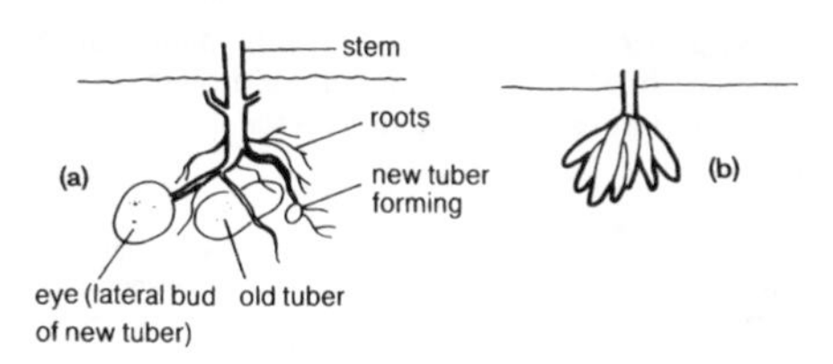

tuber (a) Stem tuber, e.g. potato. (b) Root tuber, e.g. dahlia.

Tullgren funnel An apparatus used to isolate organisms living in the air spaces in **soil**, e.g. beetles and spiders. The organisms move away from the strong light and high temperature produced by a lamp, and are collected in a jar of preservative.

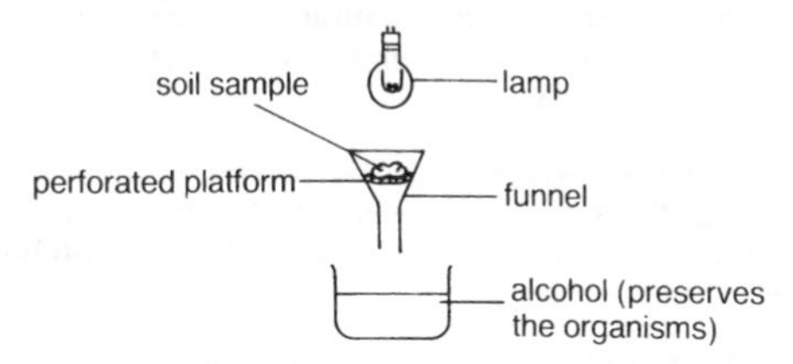

Tullgren funnel

turgor The state of a plant **cell** after maximum **water** absorption. Surrounding water enters a cell by osmosis causing the **vacuole** to expand, pushing the **cytoplasm** against the cell wall and making the plant cell solid and strong.
Turgid cells are important in supporting plants,

conferring strength and shape. Young plants depend completely on turgor for support, although in older plants, support is obtained from *wood* formation (see **xylem**).

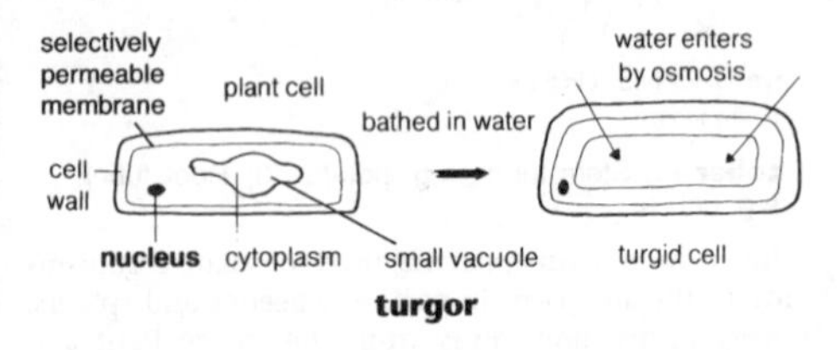

turgor

tympanic membrane See **tympanum**.

tympanum or **tympanic membrane** A thin membrane separating the outer ear and middle ear in tetrapods, i.e. the *eardrum*. See **ear**.

ulna The **posterior** of the two **bones** of the lower region of the **tetrapod** forelimb. In humans, it is the larger of the two bones of the forearm. See **endoskeleton**.

umbilical cord The cord of blood vessels linking the growing **foetus** in the womb to the **placenta**. It carries nourishment to the foetus and waste products from it. See **birth**, **pregnancy**.

unicellular (of an organism) Consisting of one **cell** only. Unicellular organisms include **protozoans**, **bacteria** and some **algae**. Compare **multicellular**.

urea The main nitrogenous excretory product of mammals. Urea is produced in the **liver** from the **deamination** of excess **amino acids**, and is then excreted by the **kidneys**.

$$H_2N-\underset{\underset{O}{\|}}{C}-NH_2$$

urea The chemical structure of urea.

ureter The tube, in vertebrates, that carries **urine** from the **kidney** to the **bladder**.

urethra The tube in mammals which conveys **urine** from the **bladder** to the exterior. In male mammals, it also serves as a channel for the exit of **spermatazoa**. See **kidney, fertilization**.

uterus or **womb** A muscular cavity in most female mammals that contains the **embryo(s)** during development. The uterus receives **ova** from the **oviduct** and connects to the exterior via the **vagina**. See **fertilization**.

urine A solution of **urea** and **mineral salts** in water produced by the mammalian **kidney**. It is stored in the **bladder** before discharge via the **urethra**.

vaccine A small quantity of **antigens** which is injected into the body. This stimulates the production of the appropriate **antibodies** against a particular **pathogen**, which are then present and available to act if and when that pathogen enters the body.

Vaccines are produced in the following ways:
(a) by separating antigens from the microorganisms (vaccine against influenza);
(b) by mass production of the antigen by **genetic engineering** (vaccine against the hepatitis B **virus**);
(c) by using the killed pathogen (vaccine against whooping cough);
(d) by chemically changing a **toxin** so that it is no longer toxic but still resembles the antigen (vaccine against tetanus);
(e) by using a live but non-pathogenic strain of the organism (vaccine against rubella).

vacuole A fluid-filled space within **cell cytoplasm**, containing many compounds such as sugars, in solution. Vacuoles are particularly important in maintaining **turgor** in plant cells. See **cell, contractile vacuole**.

vagina A duct in most female mammals which connects the **uterus** with the exterior. It receives the **penis** during **copulation**, and is the route by which the **foetus** is passed during **birth**. See **fertilization**.

valves Membranous structures within the **heart** and **veins** in animal **circulatory systems** which allow **blood** to flow in one direction only. The heart valves are the following:
(a) *mitral valve* (or *bicuspid valve*): two flaps between the left **atrium** and left **ventricle** of the heart in birds and mammals;
(b) *tricuspid valve*: three flaps between the right atrium and right ventricle of the mammalian heart;
(c) *semilunar valves*: half-moon-shaped flaps in the

mammalian heart between the right ventricle and **pulmonary artery**, and the left ventricle and **aorta**. Semilunar valves are also found in the **lymphatic system** and veins. See **heartbeat**.

variation Differences in characteristics between members of the same **species**. There are two main types:
(a) *continuous variations* in which there are degrees of variation throughout the **population** showing **normal distribution** around a mean. For example, in humans: height, weight, **pulse rate**;
(b) *discontinuous variations* are absolutely clear cut, i.e. there are no intermediate forms; for example, **blood groups** in humans. Discontinuous variations do not show normal distribution and are used when doing **genetics** crosses.

Variation within a species results either from inherited or environmental factors or a combination of both. Thus, a human being inherits **genes** influencing height, for example, but will also be subject to environmental

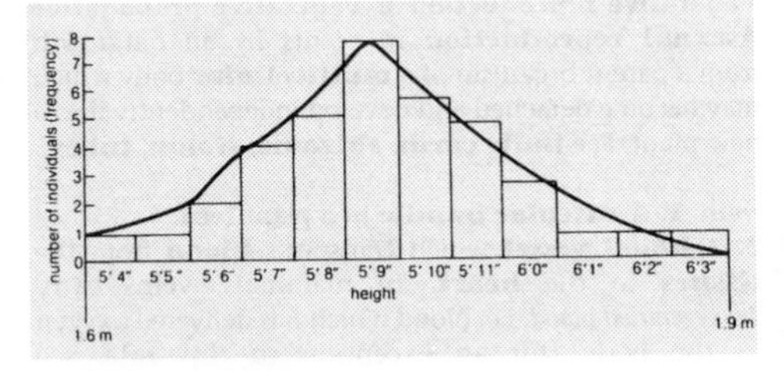

variation Height in men – an example of continuous variation.

factors such as nutrition. Inherited variations are considered to be the basis of **evolution** by **natural selection**.

vascular bundle A strand of longitudinal conducting **tissue** within plants, consisting mainly of **xylem** and **phloem**. See **root**, **stem**, **leaf**.

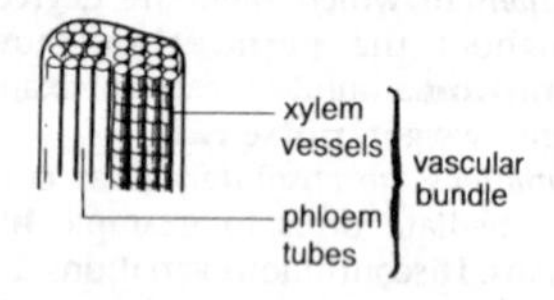

vascular bundle A vascular bundle in a stem.

vasoconstriction See **temperature regulation**.

vasodilation See **temperature regulation**.

vegetative reproduction or **vegetative propagation** **Asexual reproduction** in plants by an outgrowth from a parent organism of a **multicellular** body which may become detached and develop independently into a new plant. See **bulb**, **corm**, **rhizome**, **stolon**, **tuber**.

vein 1. A **vascular bundle** in a plant **leaf**.
2. A **blood vessel** which transports **blood** from the **tissues** to the **heart**. In mammals, veins carry *deoxygenated blood*, i.e. blood which has delivered oxygen to the body (for an exception to this rule, see **pulmonary vessels**). Veins connect with smaller vessels called *venules* which carry blood from the

capillaries. Veins are thin-walled, and since the **blood pressure** in veins is less than in **arteries**, they have **valves** to prevent the blood flowing away from the heart.

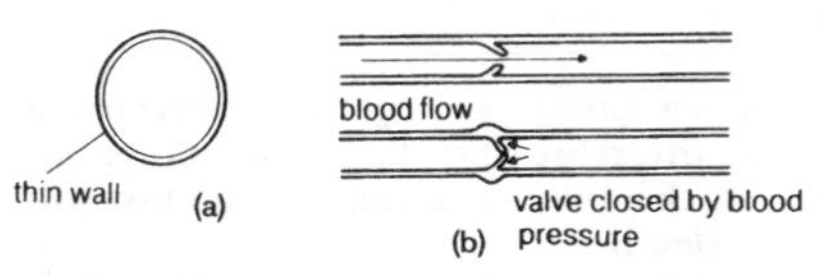

vein (a) Section through a vein. (b) Valve operation in veins.

vena cava The largest **vein** in the **circulatory system** of vertebrates. In mammals, it is either of the two main veins:
(a) *superior* (or *anterior*) *vena cava* carries **blood** from the head, neck and upper limbs into the right **atrium** of the **heart**;
(b) *inferior* (or *posterior*) *vena cava* carries **blood** from the rest of the body and lower limbs into the right atrium.

ventral Relating to features of, on, or near that surface of an organism which is normally directed downwards, although in humans, it is directed forwards. Compare **dorsal**.

ventricle See **heart**, **heartbeat**.

venule See **vein**.

vertebral column or **backbone** A series of closely

arranged bones (vertebrae) and/or **cartilages** which runs dorsally from the skull to the tail in vertebrates. It is the principal longitudinal supporting structure and encloses and protects the **spinal cord**. See **endoskeleton**.

villi (*singular* **villus**) **1.** Finger-like projections in the vertebrate **intestine** where they occur in large numbers to increase the surface area available for **absorption** of food. See **ileum**.
2. Finger-like projections which develop from the mammalian **placenta** into the **uterus** wall and thus increase the area of contact between maternal and embryonic **tissues**.

virus The smallest known living particle, having a diameter between 0.025 and 0.25 microns. Viruses are **parasites** infecting animals, plants and **bacteria**. Virus infections of man include measles, polio and influenza. A virus particle consists of a **protein** coat surrounding a length of **nucleic acid**, either **DNA** or **RNA**.

virus A virus infecting a bacterium.

viscera A collective term for the internal **organs** of an animal.

vitamins **Organic compounds** required in small quantities by living organisms. Like enzymes, vitamins play a vital role in chemical reactions within the body, often regulating an enzyme's action. Shortage of vitamins from the human diet leads to *deficiency diseases*. The properties of some important vitamins are summarized below.

Vitamin	Rich sources	Effects of deficiency
Vitamin A	milk, liver, butter, fresh vegetables	night-blindness, retarded growth
Vitamin B_1	yeast, liver	*Beri-beri*: loss of appetite and weakness
Vitamin B_2	yeast, milk	*pellagra*: skin infections, weakness, mental illness
Vitamin C	citrus fruits, fresh green vegetables	*scurvy*: bleeding gums, loose teeth, weakness
Vitamin D	eggs, cod liver oil	*rickets*: abnormal bone formation
Vitamin E	fresh green vegetables, milk	thought to affect reproductive ability
Vitamin K	fresh vegetables	blood clotting impaired

vitamins The properties of some important vitamins.

vitreous humour A transparent jelly-like material which fills the cavity behind the **lens** of the vertebrate **eye**.

voluntary muscle or **striated muscle** **Muscle** connected to the mammalian **skeleton**, and which is under the conscious control of the organism; examples are the limb muscles, muscles of face and mouth, etc. Voluntary muscles involved in limb movement are attached to **bones** by **tendons** and cause movement by contracting and thus pulling on bones, particularly at **joints**. See **involuntary muscle, antagonistic muscles.**

wall pressure The resistance to stretching in a plant **cell**. A cell absorbing water by **osmosis** will continue to expand until its **selectively permeable membrane** and cell wall can stretch no further. Wall pressure increases as the point of **turgor** is approached.

warm blooded See **endotherm.**

water A **compound** consisting of the elements hydrogen and oxygen. Its chemical formula is H_2O. Water can be synthesized by burning hydrogen in air.

$$\text{hydrogen} + \underset{\text{(oxygen)}}{\text{air}} \xrightarrow[\text{energy}]{\text{heat}} \text{water}$$

Water can be split into two **elements**:

$$\text{water} \xrightarrow[\text{energy}]{\text{electrical}} \text{hydrogen} + \text{oxygen}$$

$$\text{water} \xrightarrow[\text{energy}]{\text{light}} \text{hydrogen} + \text{oxygen}$$

Reaction (2) occurs in green plants during **photosynthesis**.
Water is a colourless, tasteless compound which freezes at 0 °C and boils at 100 °C. Water can exist in several states (the *water cycle*) as shown in the diagram.

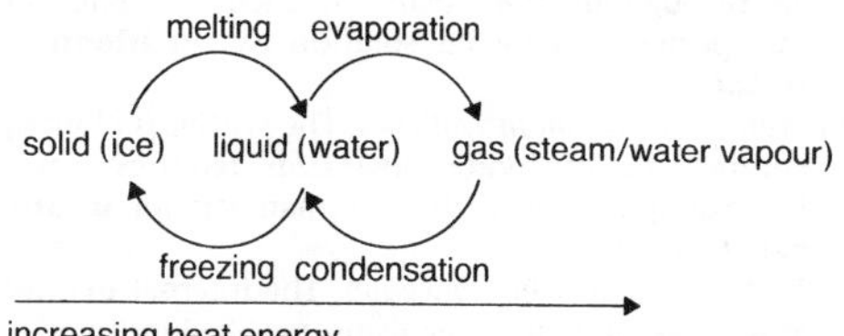

water The water cycle.

Water is biologically important for the following reasons:

(a) *Water is a major constituent of cells.* Water is the most abundant component of organisms. It is estimated that the human body is more than 60% water. The water content of living organisms can be estimated by weighing food samples, e.g. a potato, drying the samples in an oven, and then reweighing. The resulting loss of weight gives the water content.
The water content of living organisms is usually between 60% and 95%, although it can be as high as 99% (as in jellyfish) and as low as 20% (in plant seeds).

(b) *Water is a solvent.* Water is called 'the universal solvent' since more substances dissolve in water than in any other liquid. This is important, since all the chemical reactions which occur in organisms take place in *aqueous solution*, i.e. dissolved in water (*aqua* = water).

(c) *Water is a means of transport.* Within organisms, substances such as food are required to be transported throughout the organism. Such materials are transported in aqueous solution e.g. in **blood** or **phloem**.

(d) *Water is important in syntheses.* The synthesis of many compounds in living organisms requires water. For example, one of the raw materials for **photosynthesis** is water.

(e) *Water is required in lubrication.* The internal **organs** and **joints** of living organisms must be lubricated to prevent friction during movement. Various lubricating fluids exist, e.g. **mucus**, all of which are aqueous solutions.

(f) *Water is important in reproduction.* Many organisms use water to transport the male reproductive **cells** (**sperm**) to the female cells (**ovum**) so that **fertilization** can occur.

(g) *Water is important in temperature regulation.* The evaporation of water (e.g. sweat) from the surface of organisms has a cooling effect, while the high water-content of cells provides insulation and prevents rapid temperature changes.

water purification The treatment applied to tap water in order to make it safe to drink. The **water** is filtered through sand and gravel to remove large particles and then chlorine is added to kill harmful **microorganisms**.

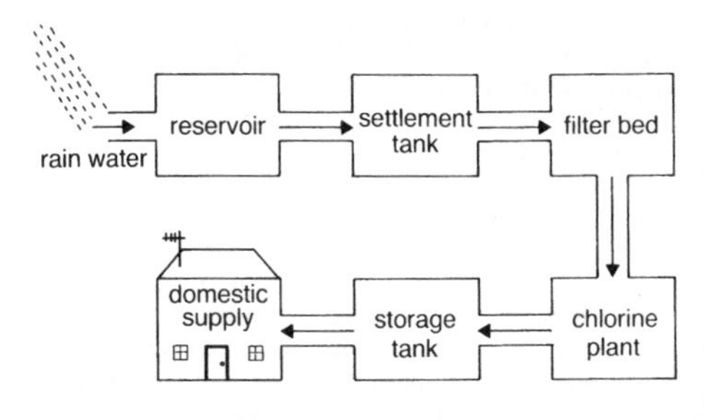

water purification

weathering The process by which exposed rock is converted into **soil**. The rock is broken down to small particles by the effects of wind, rain, heat and frost. These rock particles form the 'skeleton' of the soil but are only one component of a complex mixture of inorganic and organic factors which make up a soil.

white blood cell, white blood corpuscle or **leucocyte** One of various types of **blood cell** found in most vertebrates. Their function is in defence against **microorganism** infection, which they achieve by **phagocytosis** or by **antibody** production.

white blood corpuscle See **white blood cell.**

wild type An organism having a **phenotype** or **genotype** which is characteristic of the majority of the **species** in natural conditions.

wilting A plant condition occurring when water loss by **transpiration** exceeds water uptake. The cells lose **turgor** and the plant droops.

womb See **uterus**.

wood See **xylem**.

xylem **Tissue** within plants which conducts **water** and **mineral salts**, absorbed by **roots** from the **soil**, throughout the plant. Xylem tissue consists of long continuous tubes formed from columns of **cells** in which the horizontal cross-walls have disintegrated and the cell contents have died.
The vessels thus formed are strengthened by a

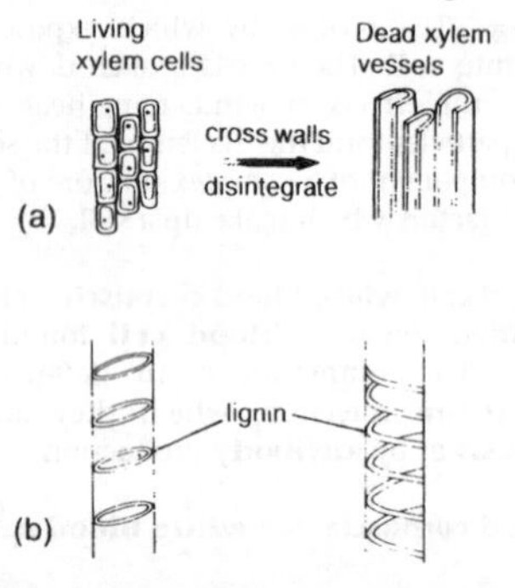

xylem (A) Long, continuous vessels are formed as horizontal cross-walls disintegrate. (b) The vessels are strengthened by lignin.

compound called **lignin**, and ultimately form the *wood* of the plant. Associated with xylem vessels, and providing additional strength, are specialized fibrous cells called *xylem fibres*, some of which are useful to humans, for example, flax. Thus xylem is commercially important as a source of wood and fibres. See **leaf**, **root**, **stem**.

yeast A **unicellular fungus** which is important in baking, **brewing** and **fermentation**. See **budding**.

yellow spot See **fovea**.

yolk A store of food material, mainly **protein** and **fat** present in the eggs (**ova**) of most animals. In fish, reptiles and birds, the yolk is contained within a *yolk sac* which is absorbed into the **embryo** as the yolk is used.

zooplankton See **plankton**.

zygote The **diploid cell** resulting from the fusion of two **gametes** during **fertilization**. The tiny ball of cells, formed by cleavage of the zygote (**mitosis**), becomes embedded in the wall of the **uterus** in mammals (**implantation**). See **fertilization**, **pregnancy**.

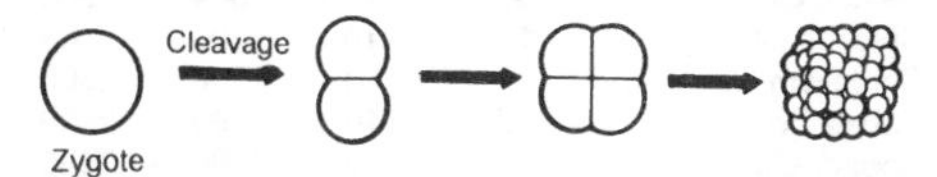

zygote The zygote divides by mitosis after fertilization, to form a ball of cells.

APPENDIX A: Chemical elements

In this table, the names and symbols of the chemical elements are given, with their proton numbers (Z), numbers of isotopes (n_i) and melting and boiling temperatures (T_m, T_b).

Element	Z	n_i	T_m/°C	T_b/°C
Actinium Ac	89	7	1230	3100
Aluminium Al	13	7	660	2400
Americium Am	95	8	1000	2600
Antimony Sb	51	18	631	1440
Argon Ar	18	7	−190	−186
Arsenic As	33	11	—	610
Astatine At	85	7	250	350
Barium Ba	56	16	710	1600
Berkelium Bk	97	6		
Beryllium Be	4	4	1280	2500
Bismuth Bi	83	12	271	1500
Boron B	5	4	2030	3700
Bromine Br	35	18	−7	58
Cadmium Cd	48	18	321	767
Caesium Cs	55	15	27	690
Calcium Ca	20	11	850	1450
Californium Cf	98	7		
Carbon C	6	6	3500	3900
Cerium Ce	58	13	804	2900
Chlorine Cl	17	10	−101	−34
Chromium Cr	24	8	1900	2600
Cobalt Co	27	10	1490	2900
Copper Cu	29	10	1080	2580
Curium Cm	96	7	1340	

Element	Z	n_i	T_m/°C	T_b/°C
Dysprosium Dy	66	12	1500	2300
Einsteinium Es	99	10		
Erbium Er	68	10	1530	2600
Europium Eu	63	12	830	1450
Fermium Fm	100	7		
Fluorine F	9	4	−220	−188
Francium Fr	87	5	30	650
Gadolinium Gd	64	14	1320	2700
Gallium Ga	31	10	30	2250
Germanium Ge	32	13	960	2850
Gold Au	79	13	1060	2660
Hafnium Hf	72	11	2000	5300
Helium He	2	3	—	−269
Holmium Ho	67	6	1500	2300
Hydrogen H	1	3	−259	−253
Indium In	49	19	160	2000
Iodine I	52	17	114	183
Iridium Ir	77	10	2440	4550
Iron Fe	26	8	1539	2800
Krypton Kr	36	19	−157	−153
Lanthanum La	57	8	920	3400
Lawrencium Lw	103	1		
Lead Pb	82	24	327	1750
Lithium Li	3	4	180	1330
Lutetium Lu	71	5	1700	3300
Magnesium Mg	12	6	650	1100
Manganese Mn	25	9	1250	2100
Mendelevium Md	101	1		
Mercury Hg	80	16	−39	357
Molybdenum Mo	42	15	2600	4600
Neodymium Nd	60	13	1020	3100
Neon Ne	10	7	−250	−246

Element	Z	n_i	T_m/°C	T_b/°C
Neptunium Np	93	8	640	3900
Nickel Ni	28	11	1450	2800
Niobium Nb	41	15	2400	5100
Nitrogen N	7	6	−210	−196
Nobelium No	102	1		
Osmium Os	76	13	3000	4600
Oxygen O	8	6	−219	−183
Palladium Pd	46	17	1550	3200
Phosphorous P	15	7	44	280
Platinum Pt	78	12	1770	3800
Plutonium Pu	94	11	640	3500
Polonium Po	84	12	250	960
Potassium K	19	8	63	760
Praseodymium Pr	59	8	930	3000
Promethium Pm	61	8	1000	1700
Protactinium Pa	91	9	1200	4000
Radium Ra	88	8	700	1140
Radon Rn	86	7	−71	−62
Rhenium Re	75	7	3180	5600
Rhodium Rh	45	14	1960	3700
Rubidium Rb	37	16	39	710
Ruthenium Ru	44	12	2300	4100
Samarium Sm	62	14	1050	1600
Scandium Sc	21	11	1400	2500
Selenium Se	34	16	220	690
Silicon Si	14	6	1410	2500
Silver Ag	47	16	960	2200
Sodium Na	11	6	98	880
Strontium Sr	38	13	77	1450
sulphur S	16	7	119	445
Tantalum Ta	73	11	3000	5500
Technetium Tc	43	14	2100	4600

Element	Z	n_i	T_m/°C	T_b/°C
Tellurium Te	52	22	450	1000
Terbium Tb	65	8	1360	2500
Thalium Tl	81	16	300	1460
Thorium Th	90	9	1700	4200
Thulium Tm	69	10	1600	2100
Tin Sn	50	21	231	2600
Titanium Ti	22	8	1680	3300
Tungsten W	74	10	3380	5500
Uranium U	92	12	1130	3800
Vanadium V	23	7	1920	3400
Xenon Xe	54	22	−111	−108
Ytterbium Yb	70	11	820	1500
Yttrium Y	39	12	1500	3000
Zinc Zn	30	13	420	907
Zirconium Zr	40	12	1850	4400

APPENDIX B: Units of measurement

Length
1 metre (m)=100 centimetres (cm)
1 centimetre=10 millimetres (mm)
1 millimetre=1000 microns (µm)
1 micron=1000 nanometres (nm)

Volume
1 litre (l)=1000 cm^3 (millilitres (ml))

Mass
1 tonne=1000 kilogrammes (kg)
1 kilogramme=1000 grammes (g)

Temperature
boiling point of water=100° Celsius (°C)
freezing point of water=0°C
normal average human body temperature=37°C

Energy
1 kilojoule (kJ)=1000 Joules (J)=240 calories (C)

Food type	Energy value
Carbohydrate	17 kJ/g
Protein	17 kJ/g
Fat	39 kJ/g

APPENDIX C: Characteristics of living things

For an organism to be considered as living it must demonstrate all of the following features:

Movement	The ability to change position either of all, or, of part of the body.
Excretion	The ability to remove from the body waste materials produced by the organism
Respiration	The ability to release energy by the breakdown of complex chemicals.
Reproduction	The ability to produce offspring.
Irritability	The ability to sense and respond to changes in the environment.
Nutrition	The ability to take in or manufacture food that can be used when required as a source of energy or as building materials.
Growth	The ability to increase in size and complexity through the production of new cell material.

APPENDIX D: The differences between plants and animals

Plants	*Animals*
Cell surrounded by cellulose cell wall	No cellulose cell wall
Large vacuoles in cells filled with cell sap	Vacuoles when present only small
Large cells with definite shape	Small irregularly shaped cells
Only restricted movement possible	Free movement possible
Response to stimulus slow	Rapid response to stimulus
Cells contain chloroplasts (chlorophyll)	No chloroplasts (chlorophyll)
Photosynthesize	Must obtain food from external sources

These characteristic differences should only be regarded as guidelines. Attempts to classify certain organisms (for example, bacteria, fungi, viruses) within these terms of reference will be difficult and have provided scientists with great problems.

APPENDIX E: The major groups of living organisms

THE ANIMAL KINGDOM (major phyla)

(a) INVERTEBRATES: Animals without a **vertebral column** (backbone)

Phylum Protozoa: Microscopic **unicellular** animals.

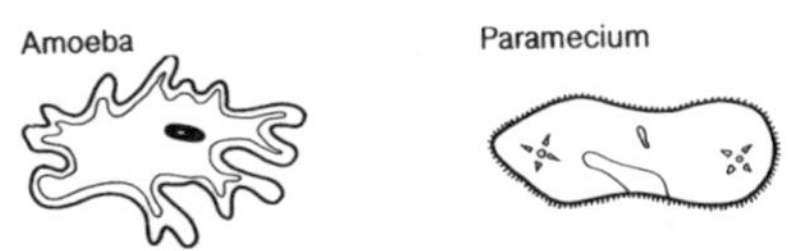

Phylum Porifera: Porous animals often occurring in colonies, for example, sponges.

Phylum coelenterata: Tentacle-bearing animals with stinging cells.

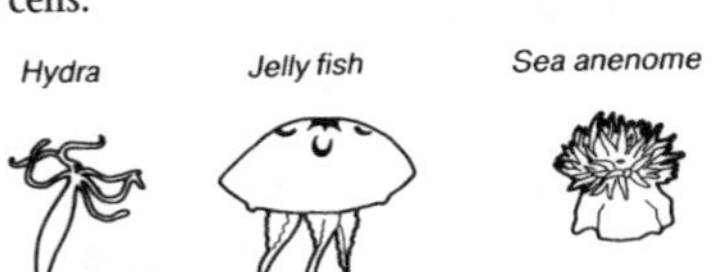

Phylum Platyhelminthes: Flatworms.

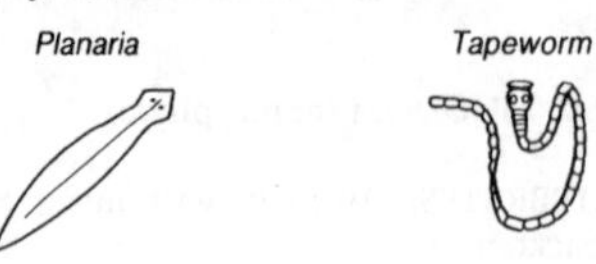

Phylum Annelida: Segmented worms.

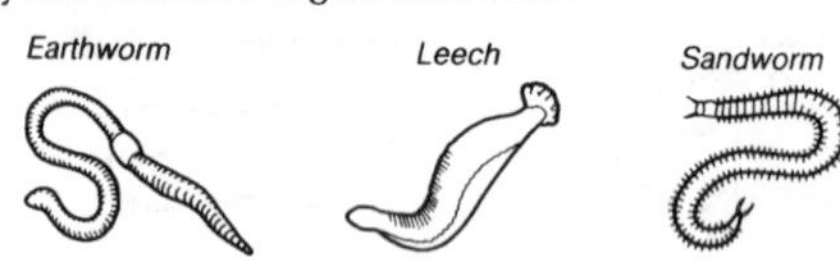

Phylum Mollusca: Soft-bodied animals often with shells.

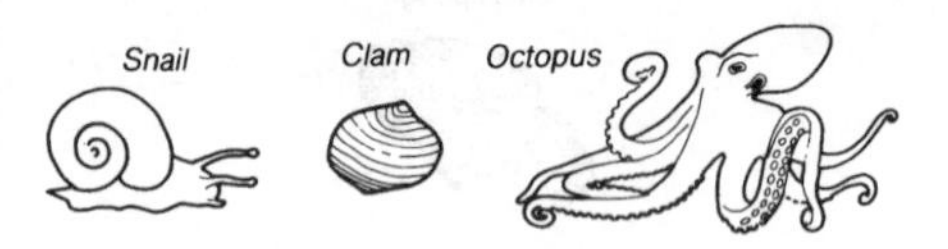

Phylum Arthropoda Jointed limbs: *exoskeleton.*

Class Insecta (*louse*) Class Crustacea (*shrimp*)

Class Arachnida (*spider*) Class Chilopoda (*centipede*)

Phylum Echinodermata: Spiny-skinned marine animals.

Starfish

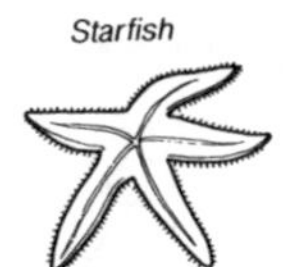

Sea urchin

Brittle star

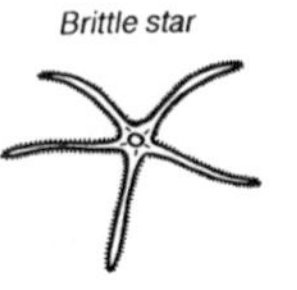

b) VERTEBRATES (Phylum Chordata): Animals with a vertebral column.

Class Pisces (Fish)
Fins, Scales, Aquatic.

Trout

Class Amphibia (amphibians)
moist, scaleless skin, live both on land and water.

Toad

Class Reptilia (reptiles) dry scaly skin.

Class Aves (birds) feathers, constant temperature.

Class Mammalia (mammals): Possess hair; maintain constant temperature; young suckled with milk.

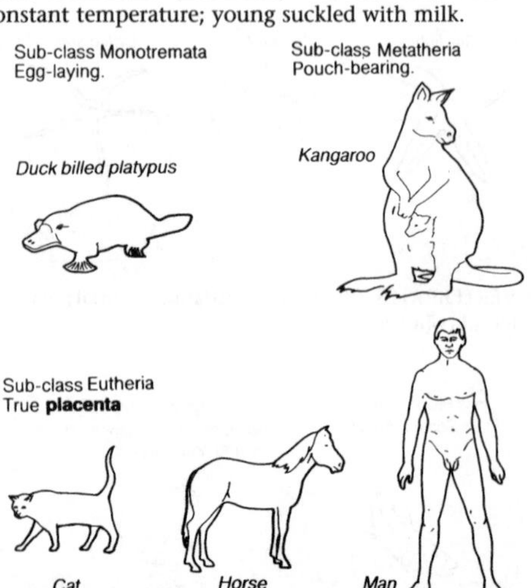

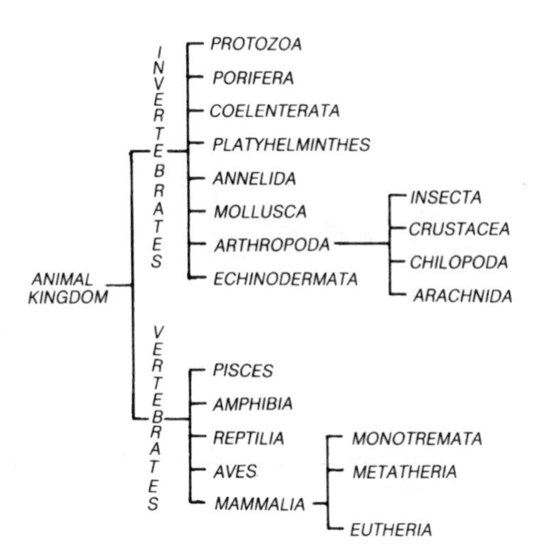
ANIMAL KINGDOM
INVERTEBRATES
PROTOZOA
PORIFERA
COELENTERATA
PLATYHELMINTHES
ANNELIDA
MOLLUSCA
ARTHROPODA
ECHINODERMATA
INSECTA
CRUSTACEA
CHILOPODA
ARACHNIDA
VERTEBRATES
PISCES
AMPHIBIA
REPTILIA
AVES
MAMMALIA
MONOTREMATA
METATHERIA
EUTHERIA

THE PLANT KINGDOM (major phyla)

Phylum Thallophyta: Unicellular and simple multicellular plants.

Class Algae: Photsynthetic; includes unicellular, filamentous, and multicellular types.

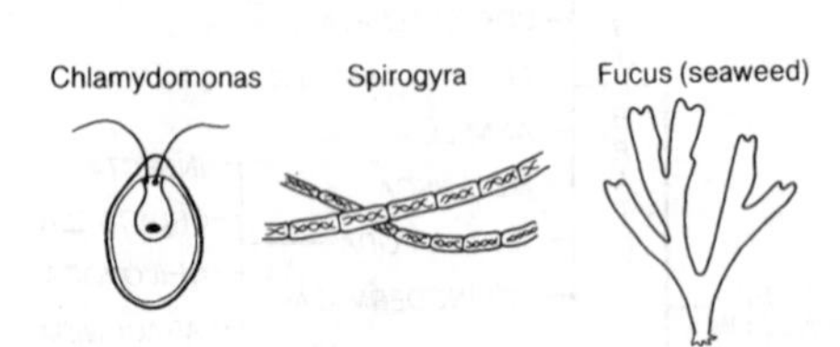

Class Fungi: **Heterotrophic**; including both **parasites** and **saprophytes**.

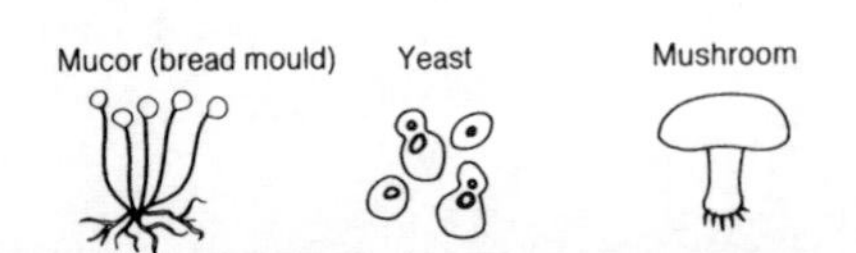

Phylum Bryophyta: Green plants with simple **leaves** and showing alternation of generations; moist **habitats**.

Class Hepaticae (liverworts)

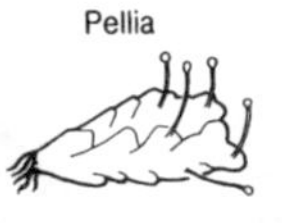

Funaria

Phylum Pteridophyta (ferns; bracken; horsetails): Green plants, with **roots**, **stems**, leaves, and showing alternation of generations.

Phylum Spermatophyta: Plants that produce **Seeds**.

Class Gymnospermae: Seeds produced in **cones**.

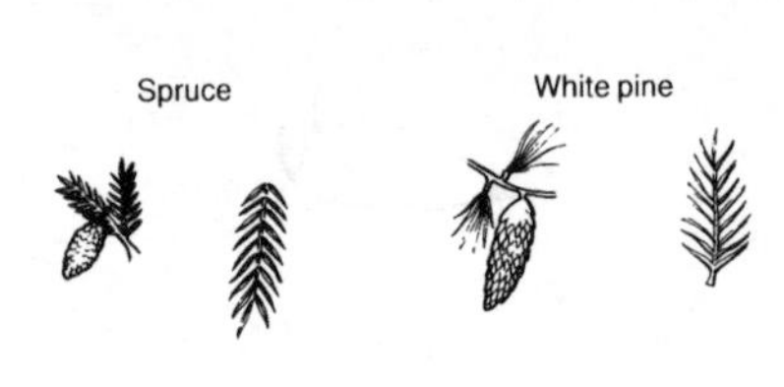

Class Angiospermae: Flowering plants; seeds enclosed within **fruits**.

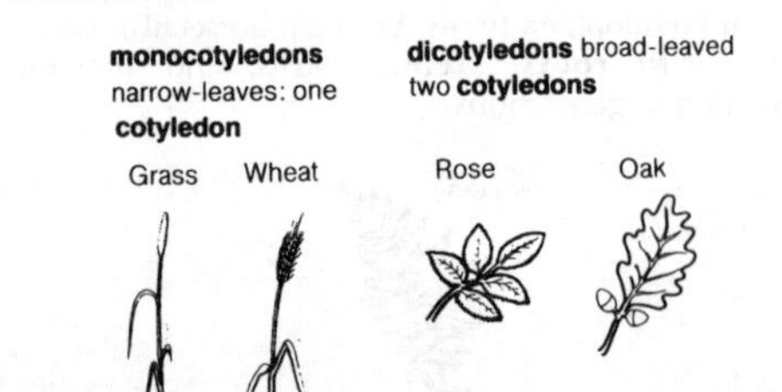

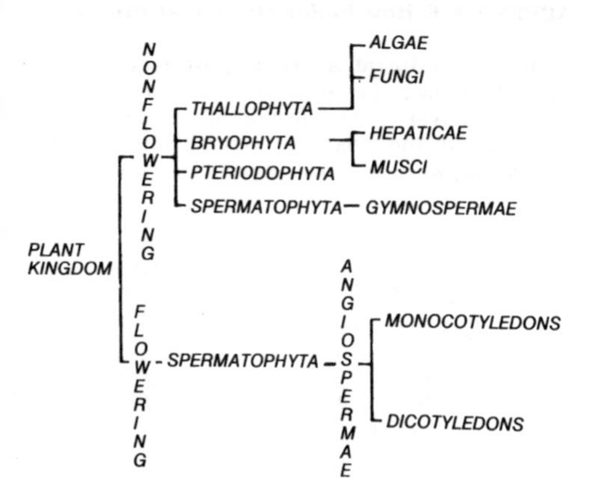

Bacteria and viruses do not meet the criteria necessary to be placed in either the animal or plant kingdoms.

APPENDIX F: How biologists look at things

Terms such as **dorsal**, **anterior**, **posterior** relate to a particular surface of an organism. The location of these areas depends on the orientation of the organism. For example, in the dog, the ventral surface points downwards, whereas in man, it points forwards.

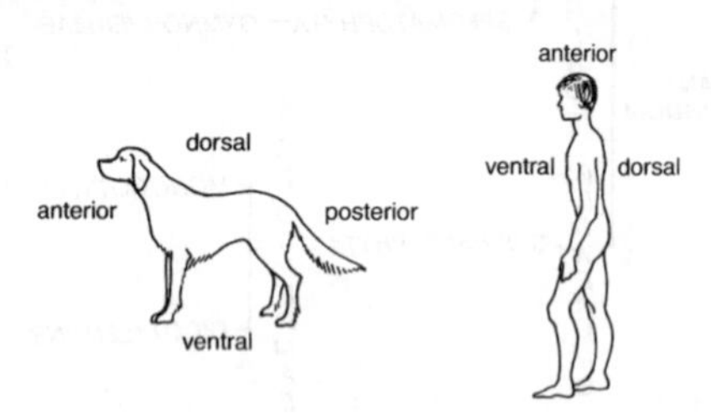

Plant sections

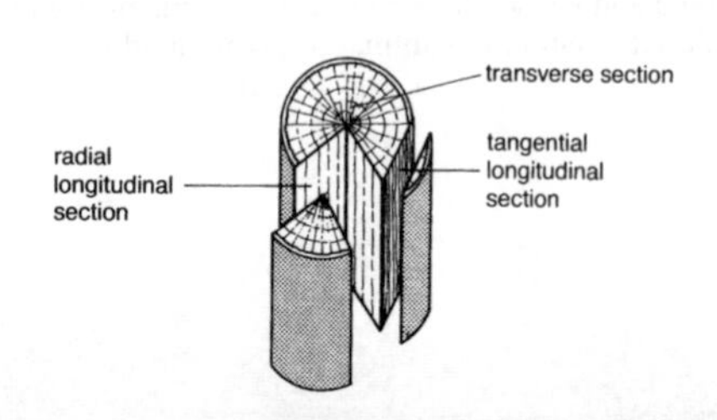

APPENDIX G: Writing up experiments

Your aim should be to give an account which would enable another scientist to copy precisely your procedure and to draw conclusions from your observations.

The traditional way of reporting an experiment includes:

Title	A summary of the aim or aims.
Introduction	A brief outline of useful background information.
Apparatus	A complete list (or labelled diagram) of everything used.
Method	An ordered account of what was done. (A labelled diagram may help.)
Result	A record of observations and measurement using tables, graphs, diagrams or description.
Discussion and Conclusion	An objective account of what has been learned, problems encountered, and further investigations that arise from the observations.

APPENDIX H: Drawing diagrams

A diagram represents something as simply as possible. You do not have to be an artist to draw one, but you do need to follow a few basic rules:

— Equip yourself with a sharp pencil, a ruler and a rubber.
— Draw diagrams large enough to show detail easily.
— Lines should be single and complete to give a neat outline.
— Lines should complete a structure without leaving holes, going too far or blocking passages.
— The use of colours or shading should only be used to make an important feature stand out.
— Diagrams require a clear title.
— Labels are used to indicate the names of the parts shown.
— Annotations are used to give a brief description of the parts shown.
— Labels and annotations should be arranged horizontally around the outside of the diagram.
— Lines (drawn with a ruler) should precisely connect the labels and annotations to the parts to which they refer.
— Label and annotation lines should never cross each other.

A good diagram is often worth more than a long description.

Appendix I: Presentation of information (pie charts, line graphs, bar charts, averages, ratios, percentages)

It is often helpful to present information in the form of a chart or graph as it makes the information easier to understand. Three common ways of doing this are shown below.

Pie chart

This is a diagram in the form of a circle (the 'pie') divided into segments ('slices') to show the proportions of the quantities to be compared.

For example:
The following information was gathered about the various eye colours of pupils in a school:

eye colour	*number of pupils*
blue	200
brown	100
grey	50
green	50

This information can be presented as a pie chart. Your pie chart must have a title and a *legend* identifying each of the segments.You convert the numerical information to visual form as follows:

The total number of pupils is 400, represented by the whole circle (100%):

200 have blue eyes and take up one half of the circle (200 is one half (or 50%) of 400)

100 have brown eyes and take up one quarter of the circle (100 is one quarter (or 25%) of 400)

50 have grey eyes and take up one eighth of the circle (50 is one eighth (or 12.5%) of 400)

50 have green eyes and take up one eighth of the circle (50 is one eighth (or 12.5%) of 400).

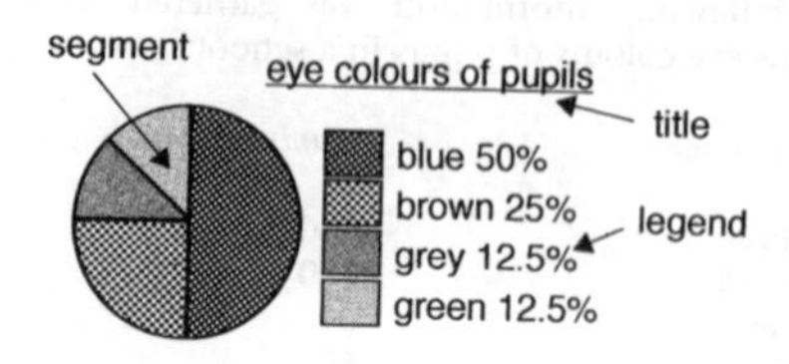

Line graph

Where two quantities are measured or counted in relation to one another, a line graph is used. For example, the following temperatures and times were noted using a thermometer and stopclock as some water cooled down in a plastic beaker:

time (min	*temperature (°C)*
0	100
10	70
20	50
30	40
40	35

These numerical measurements can be shown on a line graph. Remember to give your line graph a title. Each axis must also be given a scale and a label, and the units of measurement must be shown.

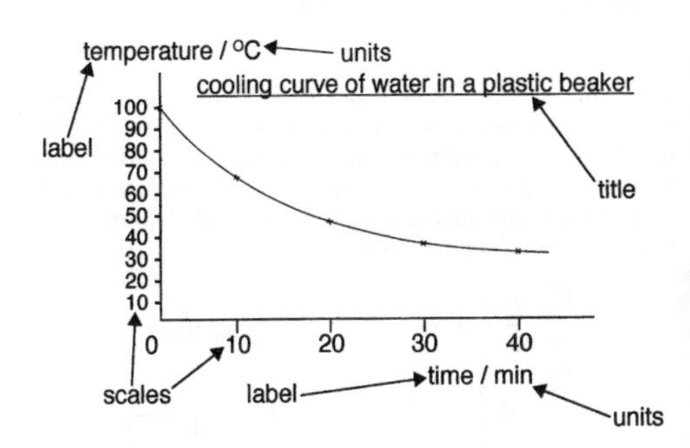

Bar charts

The bar chart, like the pie chart, can be used to show relative quantities. For example, the following information was gathered about the trees in a Forestry Commission wood in Scotland:

type of tree	*number of trees*
spruce	90
fir	55
birch	70
rowan	30
ash	50

This numerical information can be converted into separate bars, the different heights corresponding to the numbers on a vertical scale. Your bar chart must therefore be given a scale, which must be identified with a label. Each bar must also have its own label. Remember to give your bar chart a title.

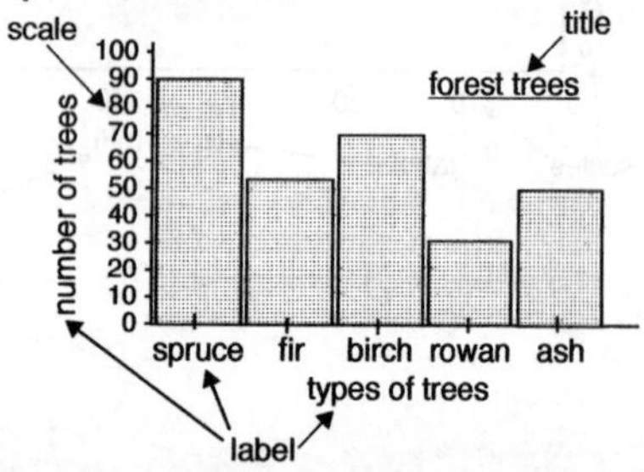

Averages

A scientist often repeats an experiment, or takes several measurements of the same thing. This is to make sure that the results can be trusted. If you have several sets of results, they are not likely to be all the same. Rather than choose just one of these results, it is fairer to take an average of the set of results.

It is not difficult to work out an average. There are two steps:

1. add up all the numbers in the set of results, to get a total

2. divide this total by the number of measurements or results in the set.

For example: find the average of the following times: 9, 11, 15, 23, 32 minutes

1. find the total:

9

11

15

23

32

---

90 minutes

2. divide the total by the number of measurements (in this case, 5):

$$90 \div 5 = 18$$

Therefore the average time is 18 minutes.

Ratios

Relative proportions can be expressed numerically as a ratio. The ratio of *a* to *b*, where *a* and *b* represent whole numbers, is

$$a \div b \text{ or } {}^{a}/_{b}$$

and is often written *a*:*b*.

For example, if the number of pupils in a biology class is 20, and 2 are absent, the the ratio of the number absent to the number present is

$$1 \div 9 \text{ or } {}^{1}/_{9} \text{ or } 1{:}9$$

Normally, ratios are expressed in the simplest whole numbers. For example:

A food sample was tested and found to contain 4 g of fat and 28 g of carbohydrate. Calculate the ratio of carbohydrate to fat in simple whole numbers.

28 g carbohydrate:4 g fat

28:4

7:1

Percentages

It is often very useful to express fractions so that their denominators are 100. A fraction with a denominator of 100 is called a *percentage*: $^{75}/_{100}$ = 75 per cent = 75%

Per cent means per hundred.

To express a fraction as a percentage:

If the total mass of the food sample above is 50 g, the percentage of fat is $^{4}/_{50} \times 100 = 8\%$

and the percentage of carbohydrate is $^{28}/_{50} \times 100 = 56\%$.

Appendix J Designing and carrying out investigations

Investigatory experimental work is a fundamental aspect of all biology courses. You will be assessed in three skills:

1. designing investigations

2. carrying out investigations

3. interpreting investigations.

When doing your investigations, use the following guidelines, which show the ten levels of attainment which are normally assessed.

Level of Attainment	Designing	Carrying Out	Interpreting
1	Ask questions abourt familiar events	Note observable features of familiar events	Discuss your observations
2	Ask questions'how?' 'why?' 'what will happen if?' Suggest ideas. Make predictions	Make a series of related observations	Use your observations to support conclusions. Compare your observations with what you expected
3	Suggest ideas, ask questions and make predictions about an event and show how they could be tested	Make careful measurements and observations using instruments	Be aware that your conclusions may be invalid unless a fair test has been done. Give a simple explanation of your observations
4	Suggest ideas, ask questions and make predictions that can be tested, based on some prior knowledge	Carry out a fair test, using suitable apparatus, and take appropriate measurements, e.g. temperatures; volumes	Identify any pattern in your results. Draw conclusions which relate the pattern to your orignal predictions

Level of Attainment	Designing	Carrying Out	Interpreting
5	Formulate a hypothesis which links the independent and dependent variables, based on scientific knowledge. Suggest a method to test your hypothesis.	Carry out a fair test, taking a range of measurements which produce results to test the relationship between the variables	Draw conclusions which link the independent and dependent variables. Consider the validity of your conclusions.
6	Suggest a relationship between two continuous variables using scientific knowledge, and how you would investigate that relationship	Carry out observations and measurements which would give fine discrimination, using appropriate apparatus	Interpret your results to explain the relationship between the two continuous variables, in terms of a suitable model
7	Predict the relative effects of two or more variables under investigation. Using scientific knowledge, suggest how to investigate your predictions	Carry out an investigation in which you control the appropriate variables, and measure their relative effects	Interpret your results in terms of the relative effects of the independent variables. Be aware of, and explain, the limitations of your evidence

Level of Attainment	Designing	Carrying Out	Interpreting
8	Generate quantitative predictions using scientific knowledge, and plan an investigation to investigate your predictions	Carry out an investigation using measuring instruments which will provide very accurate quantitative results	Consider to what extent your results support your predictions, and evaluate the contribution of each part of the investigation to the overall conclusion
9	Generate quantitative predictions, and form a hypothesis, based on a scientific theory. Suggest an appropriate range of investigatory techniques	Systematically use a range of practical investigatory techniques to obtain sufficient data to judge the relative effects of the factors involved in your predictions	Analyse and interpret your data in terms of your quantitative predictions and be aware of the uncertainty of the evidence collected

Level of Attainment	Designing	Carrying Out	Interpreting
10	Develop hypotheses to explain events you have studied using scientific knowledge, laws, theories and models. Plan extended practical investigation to test your hypotheses.	Use a range of investigatory techniques to collect reliable and valid data which will allow you to make a critical evaluation of the law, theory or model being tested	Analyse and use your data to evaluate the law, theory or model in terms of how it can explain the observed events

ILLUSTRATED BY *Ian McCullough*

CHRONICLE BOOKS
SAN FRANCISCO

First published in 1992 by
The Appletree Press Ltd
19-21 Alfred Street
Belfast BT2 8DL
Tel. +44 1232 243074
Fax +44 1232 246756

A Little Irish Songbook

First published in the United States in 1992
by Chronicle Books, 85 Second Street,
San Francisco, California 94105

Web Site: www.chronbooks.com

Distributed in Canada by
Raincoast Books
8680 Cambie Street
Vancouver, B.C. V6P 6M9

ISBN 0-8118-1535-8

9 8 7 6 5 4 3 2 1

Contents

Danny Boy

G7 C C7 F
Oh Dan - ny Boy the pipes the pipes are call - ing
C Am Dm
from glen to glen and down the moun - tain side
G7 C C7 F
The sum - mer's gone and all the ros - es fall - ing
C Dm F C
'tis you, 'tis you must go and I must bide.

And when ye come and all the flowers are dying,
If I am dead, as dead I well may be,
You'll come and find the place where I am lying,
And kneel and say an Ave there for me.
And I shall hear tho' soft you tread above me,
And all my grave will warmer, sweeter be
If you will bend and tell me that you love me,
Then I shall sleep in peace until you come to me.

The Lark in the Morning

Oh, Roger the ploughboy, he is a dashing blade,
He goes whistling and singing over yonder green glade,
He met with pretty Susan, she's handsome I declare,
She is far more enticing than the birds in the air.

Chorus

One evening coming home from the rakes of the town,
The meadows being all green and the grass it being cut down,
If I should chance to tumble all in the new mown hay,
Oh, it's kiss me now or never, love, this bonny lass did say.

Chorus

When twenty long weeks they were over and were past,
Her mammy chanced to notice she'd thickened round the waist,
It was the handsome ploughboy the maiden she did say,
For he caused me for to tumble all in the new-mown hay.

Chorus

Here's a health to young ploughboys, wherever you may be,
That likes to have a bonny lass a-sitting on his knee
With a jug of good strong porter, you'll whistle and you'll sing,
For a ploughboy is as happy as a prince or a king.

Chorus

The Curragh of Kildare

The rose upon the briar by the water running clear
Brings joy to the linnet and the deer.
Their little hearts are blessed, but mine knows no rest
For my true love is absent from me.

Chorus

A livery I'll wear and I'll comb back my hair
And in velvet so green I will appear
And straight I will repair to the Curragh of Kildare
For it's there I'll find tidings of my dear.

O you that are in love, and cannot it remove
I pity the pain you do endure,
For experience lets me know that your hearts are full of woe
A woe that no mortal can cure.

Chorus

The Galway Races

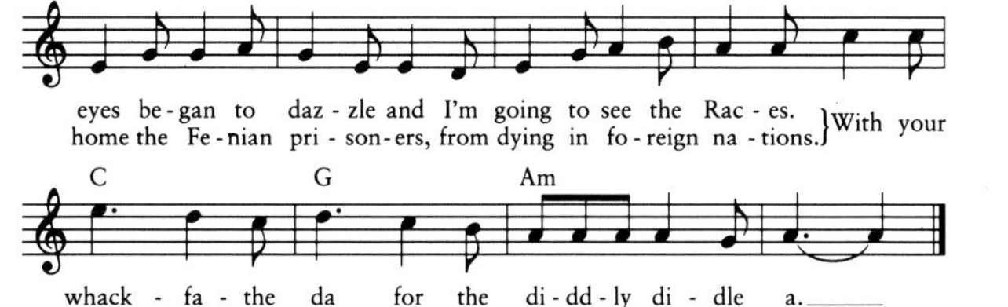

It's there you'll see the pipers and the fiddlers competing,
The nimble-footed dancers, a-tripping over the daisies.
There were others crying, cigars and likes, and bills for all the races,
With the colours of the jockeys and the price and horses' ages.
With your whack-fa-the-da, for the diddly-diddle-day.

It's there you'll see the jockeys, and they're mounted out so stately,
The pink, the blue, the orange and green, the emblem of our Nation.
When the bell was rung for starting, all the horses seemed impatient,
I thought they never stood on ground, their speed was so amazing.
With your whack-fa-the-da, for the diddly-diddle-day.

There was half a million people there, from all denominations,
The Catholic, the Protestant, the Jew and Presbyterian.
There was yet no animosity, no matter what persuasion,
But sportsman hospitality and induce fresh acquaintance.
With your whack-fa-the-da, for the diddly-diddle-day.

Believe Me, If All Those Endearing Young Charms

It is not while beauty and youth are thine own,
And thy cheeks unprofan'd by a tear,
That the fervour and faith of a soul can be known,
To which time will but make thee more dear.
No, the heart that has truly lov'd, never forgets,
But as truly loves on to the close,
As the sunflower turns on her god, when he sets,
The same look which she turn'd when he rose.

The Shores of Amerikay

It's not for the want of employment I'm going,
It's not for the love of fame,
That fortune bright, may shine over me
And give me a glorious name.
It's not for the want of employment I'm going
O'er the weary and stormy sea,
But to seek a home for my own true love,
On the shores of Amerikay.

And when I am bidding my last farewell
The tears like rain will blind,
To think of my friends in my own native land,
And the home I'm leaving behind.
But if I'm to die in a foreign land
And be buried so far far away
No fond mother's tears will be shed o'er my grave,
On the shores of Amerikay.

The Black Velvet Band

As I went walking down Broadway,
Not intending to stay very long,
I met with a frolicsome damsel,
As she came a-tripping along.
She was both fair and handsome,
Her neck it was white as a swan,
And her hair hung over her shoulder,
Tied up with a black velvet band.

Chorus

I took a stroll with this pretty fair maid
When a gentleman passed us by,
I knew she had the taking of him
By the look in her roguish black eye.
A gold watch she took from his pocket
And put it right into my hand
On the very first day that I met her,
Bad luck to the black velvet band.

Chorus

Before judge and jury next morning
Both of us had to appear
The judge he said to me, 'Young man,
Your case is proven clear.'
Seven long years' transportation,
Right on down to Van Dieman's Land
Far away from my friends and relations
Betrayed by the black velvet band.

Chorus

Carrickfergus

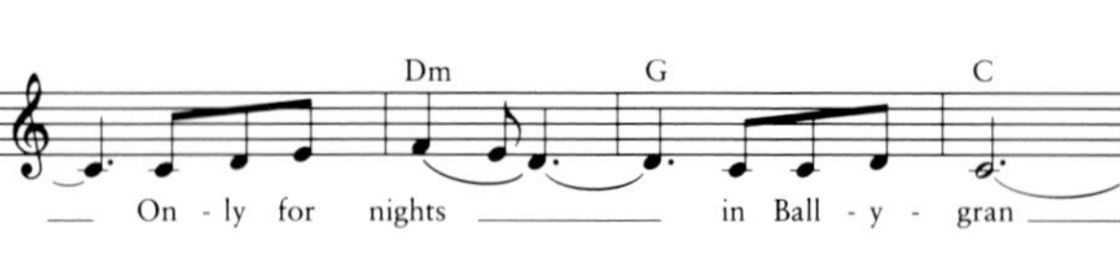

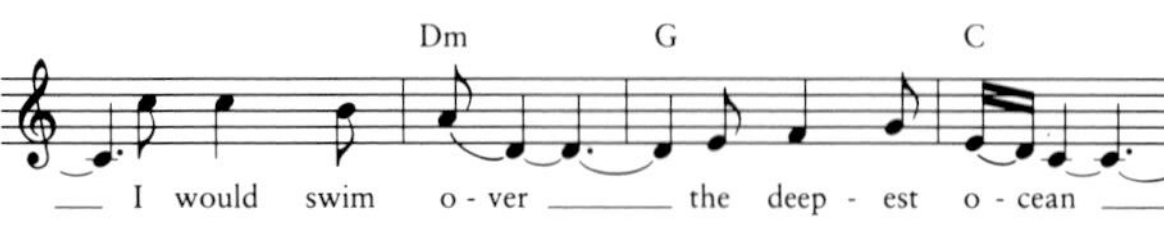

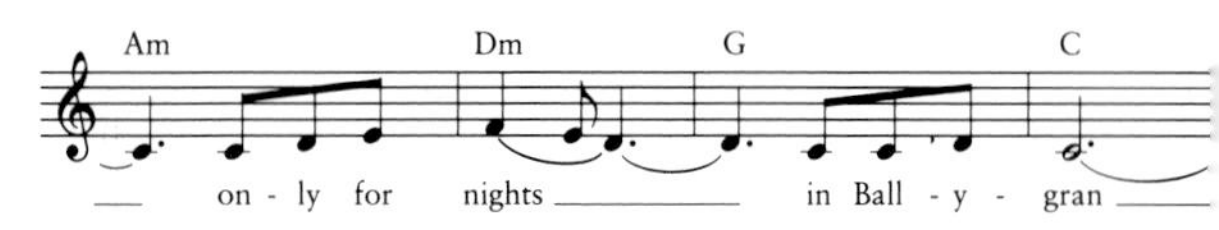

My boyhood days bring back sad reflections
Of happy hours I spent so long ago,
Of boyhood friends and my own relations
Are all passed on now, like drifting snow;
But I'll spend my days an endless rover,
Soft is the grass I walk, my bed is free;
Ah, to be back in Carrickfergus
On that long road, down to the sea.

But in Kilkenny it is reported
There are marble stones there, as black as ink,
With gold and silver I would support her
But I'll sing no more now till I get a drink.
I'm drunk today and I'm seldom sober,
A handsome rover from town to town,
Ah, but I'm sick now and my days are numbered
So come all you young men and lay me down.

I'm A Rover

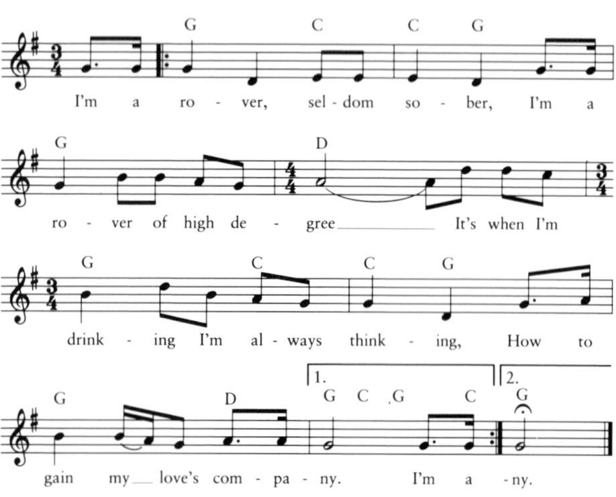
G C C G
I'm a ro - ver, sel - dom so - ber, I'm a
G D
ro - ver of high de - gree It's when I'm
G C C G
drink - ing I'm al - ways think - ing, How to
1. 2.
G D G C G C G
gain my love's com - pa - ny. I'm a - ny.

Though the night be as dark as dungeon,
Not a star to be seen above,
I will be guided without a stumble
Into the arms of my own true love.

He stepped up to her bedroom window;
Kneeling gently upon a stone
He rapped at her bedroom window:
'Darling dear, do you lie alone.

'It's only me your own true lover;
Open the door and let me in;
For I have come on a long journey,
And I'm near drenched unto the skin.'

She opened the door with the greatest pleasure,
She opened the door and she let him in;
They both took hands and embraced each other;
Until the morning they lay as one.

The cocks were crowing, the birds were singing,
The burns they ran free about the brae;
'Remember lass I'm a ploughman's laddie,
And the farmer I must obey.

'Now my love I must go and leave thee;
And though the hills they are high above,
I will climb them with greater pleasure,
Since I've been in your arms my love.'

Slieve Gallion Braes

bon - ny, bon - ny, Slieve - Gal - lion Braes.

It's oft I did ramble with my dog and my gun,
I roamed through the glens for joy and for fun,
But those days are now all over and I can no longer stay,
So farewell unto ye, bonny, bonny Slieve Gallion braes.

How oft of an evening and the sun in the west,
I roved hand in hand with the one I loved best:
But the hopes of youth are vanished and now I'm far away,
So farewell unto ye, bonny, bonny Slieve Gallion braes.

O! it was not for the want of employment at home,
That caused the young sons of old Ireland to roam,
But the rents are getting higher and I can no longer stay,
So farewell unto ye, bonny, bonny Slieve Gallion braes.

Whiskey in the Jar

I went unto my chamber all for to take a slumber,
I dreamt of gold and jewels and sure it was no wonder,
But Jenny drew my charges and she filled them up with water,
And she sent for Captain Farrell, to be ready for the slaughter.

Chorus

And 'twas early in the morning before I rose to travel,
Up comes a band of footmen and likewise Captain Farrell;
I then produced my pistol, for she stole away my rapier
But I couldn't shoot the water so a prisoner I was taken.

Chorus

And if anyone can aid me, 'tis my brother in the army,
If I could learn his station in Cork or in Killarney,
And if he'd come and join me we'd go roving in Kilkenny,
I'll engage he'd treat me fairer than my darling sporting Jenny.

Chorus

There's some take delight in the hurling and the bowling,
Others take delight in the carriages a-rolling;
But I take delight in the juice of the barley,
And courting pretty women when the sun is rising early.

Chorus

The West's Awake

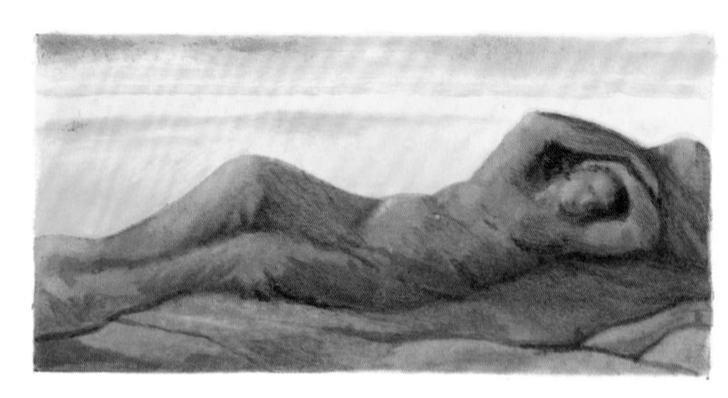

That chainless wave and lovely land
Freedom and Nationhood demand
Be sure the great God never planned,
For slumb'ring slaves a home so grand,
And long a proud and haughty race
Honour'd and sentinell'd the place
Sing, oh! not e'en their sons' disgrace,
Can quite destroy their glory's trace.

For often in O'Connor's van
To triumph dashed each Connacht clan,
And fleet as deer the Normans ran
Through Curlieu's Pass and Ardrahan;
And later times saw deeds as brave,
And glory guards Clanricarde's grave;
Sing, oh! they died their land to save,
At Aughrim's slopes and Shannon's wave.

And if, when all a vigil keep,
The West's asleep, the West's asleep,
Alas! and well may Erin weep,
That Connacht lies in slumber deep;
But hark! a voice like thunder spake;
The West's awake! the West's awake!
Sing oh, hurrah! let England quake!
We'll watch till death for Erin's sake.

The Bard of Armagh

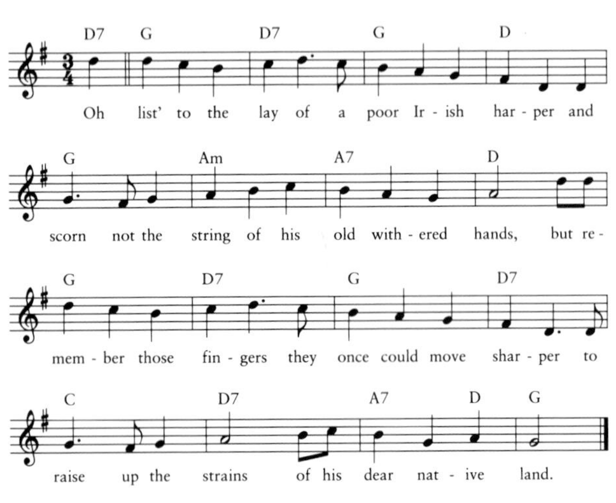

It was long before the shamrock, dear isle's lovely emblem,
Was crushed in its beauty by the Saxon's lion paw;
And all the pretty colleens around me would gather,
Call me their bold Phelim Brady, the Bard of Armagh.

How I love to muse on the days of my boyhood,
Though four score and three years have fled by them;
It's king's sweet reflection that every young joy,
For the merry-hearted boys make the best of old men.

At a fair or a wake I would twist my shillelah,
And trip through a dance with my brogues tied with straw;
There all the pretty maidens around me gather,
Call me their bold Phelim Brady, the Bard of Armagh.

In truth I have wandered this wide world over,
Yet Ireland's my home and a dwelling for me;
And, oh, let the turf that my old bones shall cover,
Be cut from the land that is trod by the free.

And when Sergeant Death in his cold arms doth embrace,
And lull me to sleep with old Erin-go-bragh,
Be the side of my Kathleen, my dear pride, oh, place me,
Then forget Phelim Brady, the Bard of Armagh.

My Singing Bird

If I could lure my singing bird from its own cosy nest,
If I could lure my singing bird I would warm it on my breast
And on my heart my singing bird would sing itself to rest,
Ah—, would sing itself to rest.

Follow Me Up To Carlow

Lift, Mac Cath - air Óg, your face, Broo - ding o'er the old dis - grace, That
Grey said vic - to - ry was sure, Soon the fire - brand he'd se - cure, Un-

Black Fitz - wil - liam stormed your place And drove you to the fern ___
-til he met at Glen - ma - lure, with Feagh Mac Hugh O' Byrne! ___

Chorus:

Curse and swear Lord Kil - dare, Feagh will do what Feagh will dare,

G D Em

Now Fitz - wil - liam have a care, Fal - len is ___ your star low.

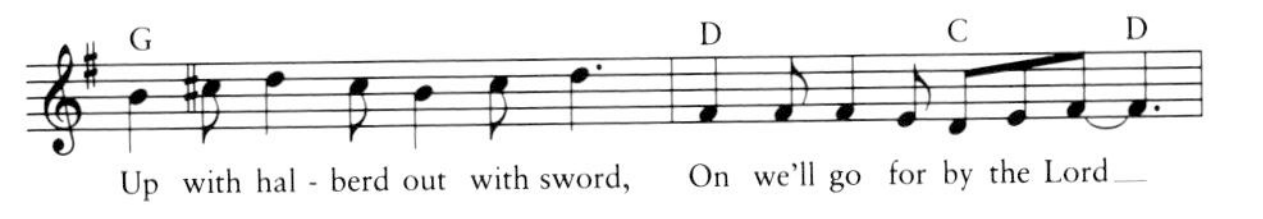

Em G D Em

Feagh Mac Hugh has giv - en the word, 'Fol - low me up to Car - low.'

See the swords of Glen Imail
Flashing o'er the English Pale,
See the children of the Gael
Beneath O'Byrne's banners.
Rooster of a fighting stock,
Would you let a Saxon cock
Crow out upon an Irish rock;
Fly up and teach him manners.

Chorus

From Tassagart to Clonmore
Flows a stream of Saxon gore,
Och! great was Ruari Óg O'More
At sending loons to Hades.
White is sick and Lane is fled–
Now for black Fitzwilliam's head–
We'll send it over dripping red
To Liza and her ladies.

Chorus

Clare's Dragoons

Another Clare is here to lead, the worthy son of such a breed,
The French expect some famous deed when Clare leads on his warriors.
Our Colonel comes from Brian's race, his wounds are in his breast and face,
The gap of danger's still his place, the foremost of his squadron.

Chorus

Oh, comrades, think how Ireland pines for exiled lords and rifled shrines,
Her dearest hope the ordered lines and bursting charge of Clare's men.
Then fling your green flag to the sky, be Limerick your battle cry,
And charge till blood floats fetlock high around the track of Clare's men.

Chorus

Pretty Susan, the Pride of Kildare

Her keen eyes did glitter like bright stars by night,
And the robes she was wearing were costly and white,
Her bare neck was shaded with her long raven hair,
And they call her pretty Susan, the pride of Kildare.

Sometimes I am jovial, sometimes I am sad,
Since my love she is courted by some other lad,
But since we're at a distance, no more I'll despair,
So my blessings on Susan, the pride of Kildare.

The Little Beggarman

I slept last night in a barn at Curraghbawn,
A wet night came on and I skipped through the door,
Holes in my shoes and my toes peeping through,
Singin' skiddy-me-re-doodlum, for old Johnny Dhu.

I must be gettin' home for it's gettin' late at night,
The fire's all raked and there isn't any light.
An' now you've heard me story of the ould ringadoo,
It's goodnight and God bless you from ould Johnny Dhu.

The Wild Rover

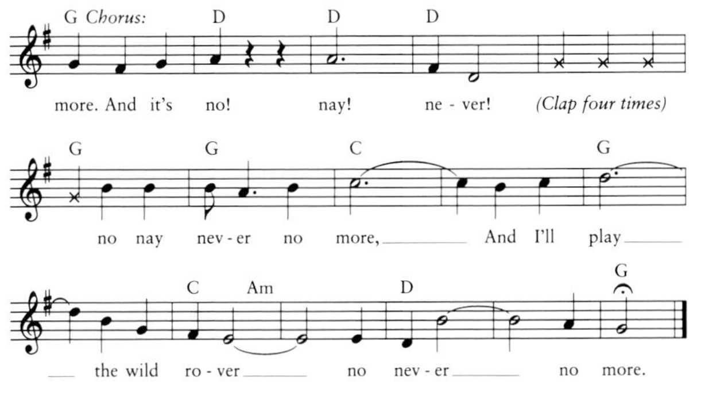

I went into an alehouse I used to frequent,
And I told the landlady my money was spent,
I asked her for credit, she answered me nay,
Saying custom like yours I can have any day.

Chorus

I took from my pocket ten sovereigns bright,
And the landlady's eyes opened wide with delight,
She said I have whiskeys and wines of the best,
And the words that I told you were only in jest.

Chorus

I'll go home to my parents, confess what I've done
And I'll ask them to pardon their prodigal son.
And when they have kissed me as oft-times before,
I never will play the wild rover no more.

Chorus

The Jolly Beggar

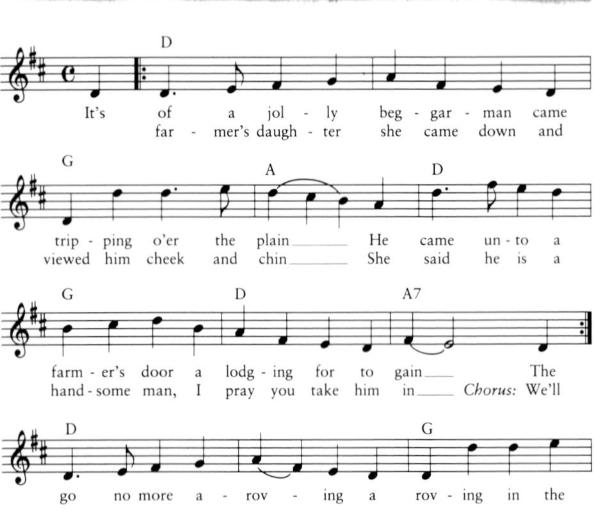

He would not lie within the barn nor yet within the byer,
But he would in the corner lie, down by the kitchen fire,
And when the beggar's bed was made of good clean sheets and hay,
Down beside the kitchen fire the jolly beggar lay.

Chorus

The farmer's daughter she came down to bolt the kitchen door,
And there she saw the beggar standing naked on the floor.
He took the daughter in his arms and to the bed he ran
Kind sir, she says, be easy now, you'll waken our good man.

Chorus

Now you are no beggar, you are some gentleman,
For you have stole my maidenhead and I am quite undone.
I am no lord, I am no squire, of beggars I be one,
And beggars they be robbers all, so you are quite undone.

Chorus

She took the bed in both her hands and threw it at the wall,
Saying, go you with the beggarman, my maidenhead and all.
We'll go no more a roving, a roving in the night,
We'll go no more a roving, let the moon shine so bright,
We'll go no more a roving.

I Will Walk With My Love

E♭ A♭
But this girl who has ta - ken my
D♭ E♭ A♭ D♭ E♭
bon - ny, bon - ny boy Let her make of him all that she
A♭ A♭
can, And whe - ther he loves me or
D♭ E♭ A♭ D♭ E♭ A♭
loves me not, I will walk with my love now and then.

The Last Rose of Summer

I'll not leave thee, thou lone one,
To pine on the stem,
Since the lovely are sleeping,
Go, sleep thou with them;
Thus kindly I scatter,
Thy leaves o'er the bed
Where thy mates of the garden
Lie scentless and dead.

So soon may I follow,
When friendships decay,
And from Love's shining circle
The gems drop away!
When true hearts lie wither'd
And fond ones are flown,
Oh! who would inhabit
This bleak world alone?

The Jug of Punch

What more diversion can a man desire,
Than to be seated by a snug coal fire,
Upon his knee a pretty wench,
And on the table a jug of punch.

Now when I am dead and in my grave,
No costly tombstone will I crave
Just lay me down in my native heath
With a jug of punch at my head and feet.

The Parting Glass

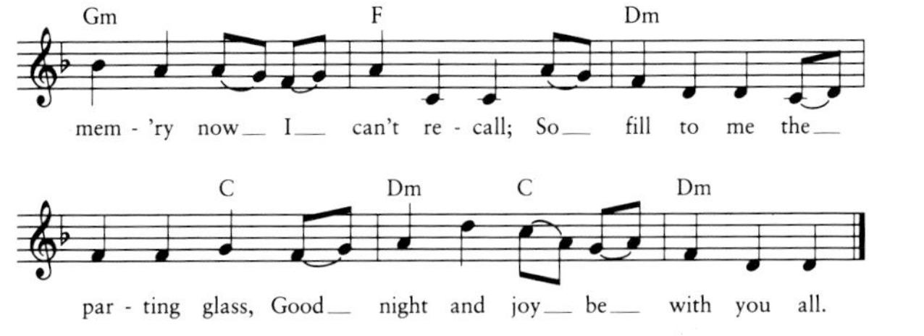

Oh, all the comrades e'er I had,
They're sorry for my going away,
And all the sweethearts e'er I had,
They'd wished me one more day to stay.
But since it falls unto my lot
That I should rise and you should not,
I gently rise and softly call,
Goodnight and joy be with you all.

If I had money enough to spend,
And leisure time to sit awhile,
There is a fair maid in this town,
That sorely has my heart beguiled,
Her rosy cheeks and ruby lips,
I own, she has my heart in thrall,
Then fill to me the parting glass,
Good night and joy be with you all.

As I Roved Out

Her boots were black and her stockings white,
And her buckles shone like silver,
And she had a dark and a rolling eye,
And her earrings tipped her shoulder.

Chorus

'What age are you, my nice sweet girl?
What age are you my honey?'
How modestly she answered me,
'I'll be sixteen age on Sunday.'

Chorus

I went to the house on the top of the hill
When the moon was shining clearly;
She arose to let me in,
For her mammy chanced to hear her.

Chorus

She caught her by the hair of the head,
And down to the room she brought her;
And with the root of a hazel twig,
She was the well-beat daughter.

Chorus

'Will you marry me now, my soldier lad?
Marry me now or never?
Will you marry me now, my soldier lad,
For you see I'm done forever?'

Chorus

'No, I won't marry you, my bonny wee girl,
I won't marry you, my honey,
For I have got a wife at home,
And how could I disown her?'

Chorus

A pint at night is my delight,
And a gallon in the morning;
The old women are my heartbreak,
But the young one is my darling.

Chorus

The Holy Ground

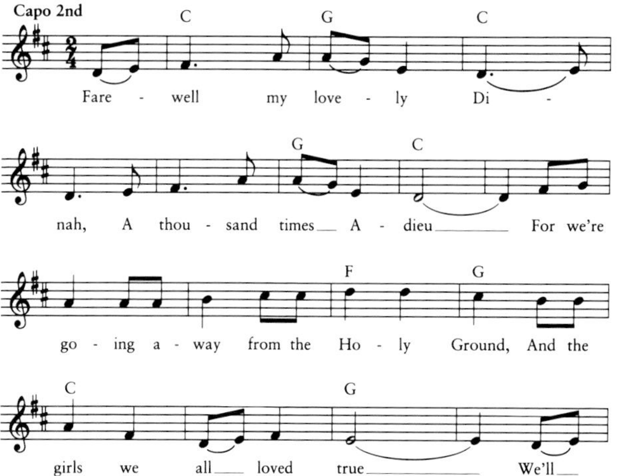

C G C
sail the South Seas o - ver and
C F G C
we'll re - turn for sure To see a -
F C
gain the girls we love and the Ho - ly
G C C
Ground once more To the girl I do a -
F G C F
dore And still I live in hope to
C G C
see the Ho - ly Ground once more.

Oh, the night was dark and stormy,
You scarce could see the moon,
And our good old ship was tossed about,
And her rigging all was torn:
With her seams agape and leaky,
With her timbers dozed and old,
And still I live in hopes to see,
The Holy Ground once more.
You're the girl I do adore
And still I live in hopes to see,
The Holy Ground once more,
Fine girl you are!

And now the storm is over,
And we are safe on shore,
Let us drink a health to the Holy Ground
And the girls that we adore;
We will drink strong ale and porter
Till we make the tap room roar
And when our money is all spent
We will go to sea once more.
You're the girl I do adore
And still I live in hopes to see,
The Holy Ground once more,
Fine girl you are!

The Kerry Dances

Was there ever a sweeter colleen in the dance than Eily More,
Or a prouder lad than Thady as he boldly took the floor.
'Lads and lassies, to your places, up the middle and down again',
And the merry-hearted laughter ringing through the happy glen.
O to think of it, O to dream of it, fills my heart with tears.

O the days of the Kerry dancing, O the ring of the piper's tune,
O for one of those hours of gladness, gone alas like our youth
too soon.